Contemporary Topics in
Analytical and
Clinical Chemistry

Volume 3

Contemporary Topics in
Analytical and Clinical Chemistry

Volume 3

Edited by
David M. Hercules
University of Pittsburgh
Pittsburgh, Pennsylvania

Gary M. Hieftje
Indiana University
Bloomington, Indiana

Lloyd R. Snyder
Technicon Instruments Corporation
Tarrytown, New York

and

Merle A. Evenson
University of Wisconsin
Madison, Wisconsin

PLENUM PRESS • NEW YORK AND LONDON

Library of Congress Cataloging in Publication Data

Main entry under title:

Contemporary topics in analytical and clinical chemistry.
   Includes bibliographical references and indexes.
   1. Chemistry, Clinical. 2. Chemistry, Analytical. I. Hercules, David M.
RB40.C66   616.07'56   77-8099
ISBN 0-306-33523-9 (v. 3)

© 1978 Plenum Press, New York
A Division of Plenum Publishing Corporation
227 West 17th Street, New York, N.Y. 10011

Printed in the United States of America

Contributors

J. A. Borders Sandia Laboratories, Albuquerque, New Mexico

Feng-Shyong S. Chuang Physical and Analytical Section, Norwich Pharmaceutical Co., Norwich, New York

J. M. Hayes Departments of Chemistry and Geology, Indiana University, Bloomington, Indiana

Gary M. Hieftje Department of Chemistry, Indiana University, Bloomington, Indiana

Gary Horlick Department of Chemistry, University of Alberta, Edmonton, Alberta, Canada

Kuang-Pang Li Department of Chemistry, University of Florida, Gainesville, Florida

David W. Peterson Departments of Chemistry and Geology, Indiana University, Bloomington, Indiana

Merle M. Millard Western Regional Research Center, Agricultural Research Service, U. S. Department of Agriculture, Berkeley, California

John P. Walters Department of Chemistry, University of Wisconsin, Madison, Wisconsin

Contents

2. Surface Analysis using Energetic Ions   57

J. A. Borders

3. A Synergic Approach to Graduate Research in Spectroscopy and Spectrochemical Analysis   91

John P. Walters

5. Signal-to-Noise Ratios in Mass Spectroscopic Ion-Current-Measurement Systems 217

David W. Peterson and J. M. Hayes

Surface Characterization of Biological Materials by X-Ray Photoelectron Spectroscopy

Merle M. Millard

1. Introduction

X-Ray photoelectron spectroscopy (XPS) is a relatively new field of electron spectroscopy. The physics of the processes involved limit the signals observed to no more than the outer 5–10 monolayers of the surface. The discrete energies of the core electrons photoejected from the elements on the surface can yield information on both elemental composition and chemical bonding.

In a relatively brief time a considerable number of experimentors have applied XPS to the structural and analytical characterization of metals, refractory oxides, and polymer surfaces.[1] The unique surface analysis capabilities of XPS have, however, been neglected as a method to study the surface of biological materials.

The combination of a surface analysis technique with a method of removing the outer surface of a specimen allows measurement of the change in concentration of various elements as a function of depth into the specimen. Auger spectroscopy and more recently XPS have been used in combination with argon ion bombardment or oxygen plasma etching (OPE) to remove outer layers of the specimen. Depth profile analysis is a powerful tool for probing the surface structure of materials and has had great success in analytical applications in the semiconductor fabrication field.[2,3]

Merle M. Millard • Western Regional Research Center, Agricultural Research Service, U.S. Department of Agriculture, Berkeley, California 94710

There is considerable current interest in cell and membrane surfaces.[4–8] This review will attempt to illustrate the usefulness of XPS for
the study of the composition and structure of surfaces of biological interest
such as cells and membranes. The application of XPS in these areas is in
its infancy. The components on the surface of cells are fairly well known.[5]
These include proteins, carbohydrates, lipids, and combinations of these
such as lipo- and glycoproteins. Less is known about their organization and
structure. The fluid mosaic model of the membrane structure is used by
most workers in the field to explain their results.[9] Answers to more
detailed problems, such as the composition of the glycoproteins in bilayer
or the mechanism of transmembrane control, are less well known. The
function of cell-surface components is quite varied and of considerable
interest in such fields as cancer research and immunology. A few selected
examples will be given to illustrate the complex and specific informational
role of components on various cell surfaces: (1) Lymphoid cells contain
antigen-recognition sites on their surfaces, and the general nature of the
immune response is the recognition of foreign proteins on the cell surface
and the production of antibodies in response to it.[10,11] (2) When tissue
culture cells become virus transformed, the glycolipid and glycoprotein
composition of the cell surface changes.[12] One demonstration of the difference in cell surface carbohydrates is the fact that transformed cells
interact with lectins or agglutinins such as concanavalin A quite differently.[5] (3) Other cell-surface receptors include those for hormones such
as insulin in liver and fat cell membranes.[13] The outer envelopes of gram-
positive bacteria contain teichoic acids which are known to be receptors for
bacteriophages and to act as antigenic determinants.[14]

It is reasonable to assume that XPS can be used as a surface-analysis
technique to study reactions at the cell surface and probe cell-surface
structure in much the same way as it has been used to study reactions at
catalytic surfaces or polymer surfaces.

2. Principles

2.1. Recent Reviews

There are a number of excellent introductory review articles dealing
with the elementary principles of XPS. Rather than duplicate this effort
the following articles can be consulted: Hercules,[15] Swartz,[16] Brundle,[17]
Clark,[18] and Siegbahn.[19] A recent book on photoelectron and Auger
spectroscopy by Carlson can be consulted for most aspects of the subject.[20]
The principles of instrumentation are discussed by Brundle,[21] and a
description of an instrument capable of high resolution has been published

in *Science*.[22] The book by Carlson[20] also contains a section on principles of instrumentation.

The performance characteristics of the XPS instruments currently available are discussed in articles by Lucchesi and Lesler[23] and Evans.[24]

Several parameters are of fundamental importance in the application of XPS to surface studies and some of these will be considered briefly.

2.2. Line Sensitivities

The relative elemental core electron line intensities are of importance in several areas of XPS. These intensities or line areas allow estimation of the surface atom concentration. This is far from straightforward and care must be exercised when attempting to relate surface composition to relative intensities or areas. The author has compared the relative elemental line intensities in proteins and chemically modified proteins with elemental composition determined by wet chemical analyses.[95,96] The agreement between the composition calculated from XPS line intensities using Wagner's sensitivity factors[25] and that found by chemical analysis was within 5%. Tables of relative core electron line intensities have been published.[25-27] Knowledge of these intensities is also quite helpful when designing experiments using XPS for surface studies.

2.3. Electron Mean Free Paths

The relative escape depth of electrons photoejected from the surface region of solids is obviously of critical importance.

The mean free paths of electrons whose kinetic energy is between 0 and 4000 eV escaping from metal and oxide films ranges between 0.5 and 4.0 nm. Typical mean free paths encountered for electrons photoejected with MgK_α or AlK_α excitation are in the range of 0.5 to 1.5 nm.[17,20] Little information exists for the mean free path of electrons from organic solids.

Clark[28] derived values for the mean free path of F 1s and 2s electrons from surface-fluorinated ethylene. A relationship between the F electron line intensities, the fluorinated film thickness, and the mean free paths was calculated. Reasonable values for some of these parameters were inserted into the equation and used to obtain limits on the range of values for others. Using this approach, values of 0.7 and 1.19 nm were derived for the mean free path of electrons from the F 1s and F 2s levels.

Andrade *et al.*[29] have recently measured the mean free paths of Au core electrons in the kinetic energy range of 900–1400 eV in gold and gold covered with barium stearate multilayers. The mean free paths in barium stearate range from 4.5 to 6.2 nm for electrons of about 900–1400 eV. The

mean free paths in gold range between 0.7 to 2.5 nm for electrons of
about 700 and 1400 eV, respectively.

Andrade reviewed the previous literature on escape depth measurements from organic solids. The mean free paths of electrons in organic
solids range between 4.0 and 15.0 nm from these studies, and it appears
that the mean free path of electrons in organic solids is greater than that
in metals and oxides.

Angular-dependent XPS measurements allow enhancement of the
extreme surface signals, and a recent review can be consulted for details.[30]

2.4. Electron Line Shapes and Spectral Deconvolution

The overlapping of lines in electron spectra is a common occurrence.
The resolution of most instruments is between 1 and 2 eV, and the linewidths observed from solids is also between 1 and 2 eV or more. It is
usually necessary to have some scheme for spectral deconvolution for the
interpretation of the spectra to advance beyond the qualitative level.

We have used a nonlinear least-squares fitting program developed by
Claudette Lederer[31] and described in C. S. Fadley's thesis.[32] This program
uses Lorentzian and Gaussian functions that are described by four independent parameters: the width, position, ratio of tail height to peak
height, and intensity factor. A linear background with a given slope may
be added to the peak shape, and the slope and intercept specify the
background.

Each curve must be approached experimentally, and there seems to
be no general rule to predict when Gaussian or Lorentzian peak shapes
will fit the spectra or when to use a leading constant or exponential tail. In
practice, the program works well and, with experience, satisfactory fits can
usually be obtained in three trials or less. The deconvoluted carbon, nitrogen, and oxygen 1s spectra from plasma-oxidized wool are good examples
of the results obtainable with the aid of this program. These spectra are
given in Section 5.

3. Specimen Preparation

3.1. Biological Samples

The biological specimen in its normal physiological state is usually not
compatible with the sample restrictions imposed by most physical methods.
The most severe restriction imposed by XPS is the requirement that the
sample be under a vacuum of approximately 10^{-6} Torr. To meet this
requirement, excluding the possibility of studying frozen samples, the

specimen must be evacuated and dehydrated. The disruptive effects of dehydration on biological samples are well known. An extensive literature exists in the field of electron microscopy (EM) concerning methods of sample preparation and dehydration and the effects of these procedures on samples as studied by EM.[104-107] References are given in Section 4.1 together with additional discussion.

The electron microprobe analyzer has been used to detect diffusible ions in cells and tissues, and quick freezing or quenching followed by vacuum dehydration seems to be the method of choice for sample preparation.[33] No systematic study has been reported dealing with the question of optimal cell and tissue specimen preparation for study by XPS.

Table 1 summarizes the XPS literature on surface analysis of biological samples with the specimen preparation indicated.

3.2. Substrates

The specimen must be mounted on some type of support that positions the sample in the spectrometer and that may be exposed to radiation. In cases where the specimen covers the support area and completely attenuates signals arising from the support, there is no interference from the support. Often, to meet requirements to preserve surface morphology or sample integrity, the substrate or support is partially exposed. In these cases, it is necessary to have a support of reproducible known composition or one that is free of elements that will interfere with the specimen. When one is interested in elements in low concentration, a clean substrate free of interferring elements is critical. We have found that substrates consisting of SiO vacuum deposited on glass disks are quite free from contamination from such elements as phosphorus, sulfur, and alkali metals.[101]

The preparation of specimen supports and the elimination of background interface problems are discussed in Sections 8 and 9. Surface coverage can be an important consideration when sorting out substrate and specimen signals. When cells or microorganisms are suspended onto supports, the substrate signal can be used to estimate surface coverage. The silicon substrate signal was used to estimate the coverage of glass disks by tissue culture cells. This point is discussed at the end of Section 9, dealing with Balb 3T3 cells.

4. Depth Profile Analysis

The concentration of a given element as a function of its depth into the surface can be of great consequence in locating structural features. In principle, one can locate regions where a given element has a high or low

Table 1. Methods of Specimen Preparation Prior to Surface Analysis by XPS

Sample	Preparation method	Result	Reference
Bacterial cell walls	Freeze drying of suspension	Mg^{+2}	Baddiley et al.[34]
Erythrocytes	Vacuum drying	Surface analysis, depth profile, analysis by argon ion etching	Meisenheimer[35]
Bacterial cells and walls	Suspension on support disk, air drying	Surface and depth profile analysis by oxygen plasma etching	Millard et al.[101]
Balb 3T3 mouse tissue culture cells	Rapid vacuum drying, glutaraldehyde fixing and critical point drying with Freon 113 in ethanol	Surface and depth profile analysis by oxygen plasma etching	Millard and Bartholomew[37]
Hepatoma cells during DNA synthesis	Air drying	Depth profile analysis of calcium in DNA synthesizing vs. quiescent cells	Pickart et al.[38]
Erythrocyte ghosts	Drying in stream of N_2, fixation with glutaraldehyde and osmium tetraoxide dehydration in graded ethanol solutions	Oxidation states of osmium	White et al.[39]

concentration and correlate this knowledge with some structural feature such as a lipid membrane bilayer or elements associated with enzyme complexes on one side of a membrane.

To accomplish depth profile analysis, a surface analysis method is combined with a method of removing known amounts of the surface. Traditionally, Auger spectroscopy and argon ion etching have been used. Depth profile analysis has found wide application in the electronics and metal industries. A general introduction to this area of analysis may be gained by reading the articles by Evans[2] or Palmberg.[3]

4.1. Argon Ion Etching

Argon ion sputter etching has a serious drawback in applications where less stable surfaces, such as organic polymers or biological specimens, are encountered. Aside from the possibility of nonuniform removal of atoms on the surface, the composition of the surface may be altered.[40] Disproportionation of oxides as a result of argon ion etching has been reported.[41,42] The disproportionation of CF_2 has been observed by XPS as a result of argon ion bombardment.[18]

4.2. Oxygen Plasma Etching

As an alternative to argon ion etching we have explored the use of oxygen plasma etching (OPE) to gently remove outer layers of biological samples. The applications of this technique for sample preparation in analytical applications and etching for fine structure studies have recently been reviewed.[43,44]

A low-temperature oxygen plasma contains strong oxidizing species such as atomic oxygen and excited molecular oxygen. These species can accomplish oxidation at ambient temperatures without disrupting surface structure. Elements forming volatile oxides are lost upon oxidation. Elements removed from the surface include carbon and nitrogen. Elements forming less volatile oxides, such as phosphorus, sulfur, and the alkali metals, are converted to nonvolatile oxides and retained on the surface as an ash.[45]

When the surface becomes covered with ash to a certain depth, the depth profile interpretation of concentration of the elements retained no longer applies. The uniformity of OPE is also a problem of concern. Electron micrographs of the etched specimen can be used to determined when nonuniform etching becomes important. The rate of OPE also must be known to construct depth profile curves.

Thomas, Millard, and Scherer[46] have estimated the rate of plasma-generated atomic oxygen etching of organic substrates to be approximately

9.0 nm/min by measuring the dimensional changes of styrene beads after exposure to oxygen plasma for varying periods of time. The actual rate of oxidation of a given substrate may differ from this rate; however, it serves as a way to calibrate the rate.

5. Review of Other XPS Studies of Systems of Biological Interest

Most of the early XPS work in areas of biological interest has been cited in two reviews.[1,47] Much of this work involves the measurement of binding energies of atoms in metallo proteins, porphyrins, and phthalocyanines.

There have been several recent XPS studies on porphyrins, metalloporphyrins, and related compounds.[48-54] Recent XPS studies on metalloproteins and enzymes include studies on superoxide dismutase cuprein,[55] copper AMP,[56] iron sulfur proteins,[57] mercury thionein,[58] glutathione peroxidase,[159,160] and hepatic zinc thionein.[61] Chlorophyll A monohydrate,[62] D-penicillamine,[63] pyrrole pigments,[64] and selenium containing amino acids[65] have also been the object of recent XPS investigations.

6. Wool Fiber Analysis

Since wool is a naturally occurring protein fiber and proteins are common components of most biological surfaces, surface studies of this fiber allow development of data interpretation and treatment techniques that are useful for biological systems such as cell surfaces. Results obtained from wool fiber surfaces are also indicative of what may be accomplished with more complex systems, i.e., those containing other components such as carbohydrates and lipids. The following will be a brief review of some recent work done in this laboratory on wool fiber analysis after various chemical modifications such as oxygen plasma oxidation or coating with fluorocarbon films.

Siognet and others have used XPS to analyze the surface of cotton textiles.[66-69]

6.1. Oxygen Plasma Surface Treatment

We have investigated the use of oxygen plasmas to effect surface changes on wool fibers and woven textiles and used XPS to study these changes.[70-73] The oxygen plasma system used for these studies is shown in Figure 1, and the apparatus and procedure is described in the literature.[74]

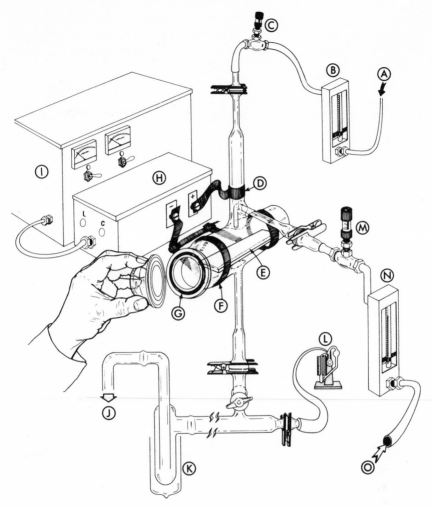

Figure 1. Flow discharge system used for the synthesis of fluorocarbon films: (A) from gas cylinder; (B) rotometer; (C) flow control valve; (D) high voltage; (E) substrate; (F) ground; (G) O-ring; (H) coupling unit; (I) generator; (J) to pump; (K) −196 °C trap; (L) McLeod gauge; (M) flow control valve; (N) rotometer; (O) from gas cylinder.[56]

Before initiating a detailed study of sample, it is often instructive to run a wide-scan survey spectrum to find out what elements are on the surface and to estimate the relative intensities of the core lines from these elements. The region between 550 eV and 50 eV contains strong lines due to most of the elements expected to be present on the surface.

A wide-scan spectrum over the region 550–50 eV taken from a sample

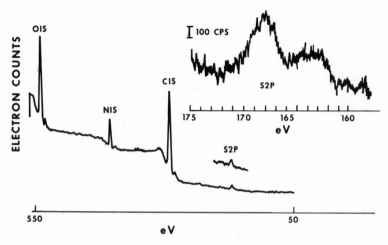

Figure 2. Wide-scan spectrum of corona-discharge-treated wool top.

of corona-discharge-treated wool top (combed, cleaned wool prepared for spinning) is shown in Figure 2. Strong lines are present at 533 eV O 1s, 400 eV N 1s, 285 eV C 1s, and weak lines due to sulfur at 227 eV S 2s and 168 eV S 2p. The S 2p spectrum contains a line at 168 eV, resulting from oxidation of disulfide or sulfhydryl bonds in the wool.

6.2. Effect of Surface Roughness on Electron Spectra

Most of the spectra are taken from woven wool swatches or wool top fibers that have been plasma treated. Woven textile or wool top fiber represents the case of an extremely rough surface. The effect of surface roughness on electron spectra is a problem of concern.

Wool textile or fiber can be pressed into a thin transparent film by applying pressure to a sample between heated plates in a hydraulic press.[75] XPS spectra were run on samples of oriented top fibers and thin flat transparent films fabricated from the fibers. The N 1s and S 2p lines are shown in Figure 3. Spectra shown in part (A) are from the oriented wool fibers, and spectra (B) are from pressed flat smooth thin films. Electron line intensities from these two samples are given in Table 2.

The changes in spectra going from an irregular surface to a smoother surface seem to be a narrowing of electron lines and an increase in line intensity by a factor of about 3. There seems to be no new information in the spectra of the thin film, and although the quality of data is improved, the results are similar. It appears that one can proceed with some confidence to study irregular surfaces and hope to obtain useful information.

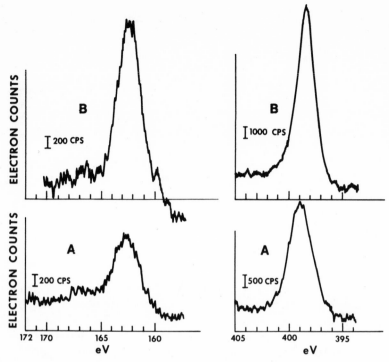

Figure 3. Sulfur 2p and nitrogen 1s electron spectra of wool as (A) oriented fibers, and (B) thin transparent films.

6.3. Spectral Deconvolution

Photoelectron spectra were obtained on a Varian IEE 15 spectrometer. Fabric samples were attached to double-backed tape and positioned on the sample mount. The binding energy region containing the desired spectra were scanned 20 or 30 times, depending on the region of interest and the intensities of the lines. The data were fitted with Lorentzian or Gaussian functions by a nonlinear least-squares fitting program.

Table 2. Electron Line Intensities from Isolated Wool Top Fibers and Thin Transparent Film Pressed from Wool Fibers

Sample	Electron line intensities (10^3 cps)			
	C 1s	O 1s	N 1s	S 2p
Unoxidized wool top fibers	30.6	12	5	1.4
Unoxidized wool top fibers pressed into a thin film	100	36.6	15.5	3.2

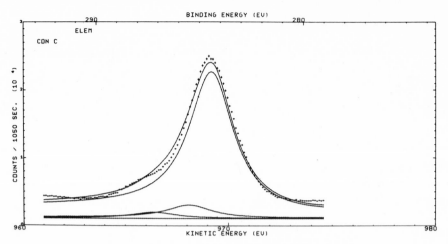

Figure 4. Carbon 1s spectrum from untreated wool top.

The program and its application for deconvolution of electron line spectra have been described in Section 2.4. Linewidths obtained from pure amino acids on the Varian instrument were about 2.5 eV wide. Carbon and O 1s electron spectra were fitted with electron lines whose linewidths at half maximum (FWHM) were 2.6 eV. Nitrogen 1s electron spectra were fitted with electron lines whose FWHM was 2.4 eV. Carbon spectra were fitted with Lorentzian functions with a leading constant tail. Oxygen and nitrogen lines were fitted with Gaussian functions with an exponential

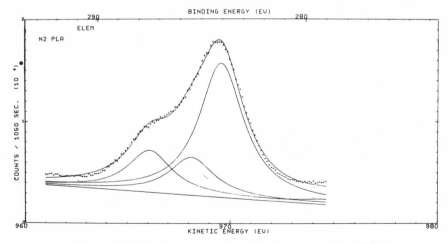

Figure 5. Carbon 1s spectrum from N_2 plasma-treated wool.[55]

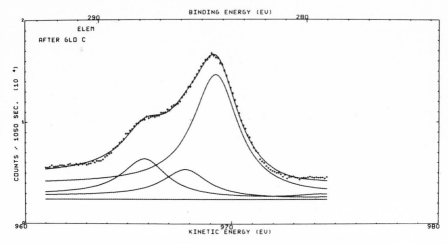

Figure 6. Carbon 1s spectrum from plasma afterglow treatment.[55]

tail. Binding energies (BE) were referenced to the carbon 1s electron line at lowest binding energy assumed to occur at 285 eV.

6.4. Carbon, Oxygen, and Nitrogen 1s Electron Spectra

The C 1s electron line spectra obtained from the control, nitrogen, and the afterglow plasma treatment are shown in Figures 4, 5, and 6. The parameters for the carbon electron lines are given in Table 3.

Table 3. Carbon 1s Electron Line Parameters, Spectra from Plasma-Treated Wool Fiber

Sample	Binding energy (eV)	Relative area	Shift in peak position (eV)
Untreated wool fiber	285.0	1.0	0.0
	286.07	0.092	1.07
	287.77	0.043	2.77
Nitrogen plasma-treated wool fiber	285.0	1.0	0.0
	286.43	0.28	1.43
	288.46	0.31	3.46
Afterglow plasma-treated wool fiber	285.0	1.0	0.0
	286.56	0.24	1.56
	288.42	0.33	3.4
Oxygen plasma-treated wool fiber	285.0	1.0	0.0
	286.55	0.23	1.55
	289.04	0.45	4.04

The C 1s electron line spectra were fitted with three lines. The position of the two high binding energy lines shifted several tenths of an electron volt going from the control spectra to the spectra from the oxygen-plasma-treated sample. The intermediate electron line shift of 1.07 eV from the 285 eV line in the control and 1.5 eV in the plasma treated samples is in the region reported by others for carbon containing a single bond to oxygen.[76] The high binding energy line shifts from 2.77 eV in the control to 4.04 eV in the spectra from the oxygen-plasma-treated sample. This electron line is in the region reported for carbonyl carbon.[76,77] Individual functional groups falling into each of the above broader structural classifications, such as alcohols and ethers or carboxylic acids and ketones, cannot be differentiated. As the surface population of carbon atoms with higher binding energies increases, the composite line at higher binding energy shifts to higher values. The obvious feature to note concerning these spectra is the increase in intensity of electron lines at higher binding energies. These lines represent oxidized carbon species and as expected, the line in the carbonyl carbon region was most intense for the oxygen-plasma-treated samples.

Oxygen 1s electron line spectra for untreated wool fiber and plasma-treated wool fiber are shown in Figures 7 and 8. Line positions and intensities are given in Table 4. The oxygen electron spectra could be resolved into two lines. The relative intensity of the higher oxygen electron line decreased by a factor of about 3 after plasma treatment. The assignment of these two oxygen lines is difficult. Very few data exist in the literature on the binding energies of structurally different oxygen atoms bound to carbon. Lindberg et al.[78] reported O 1s lines for a number of

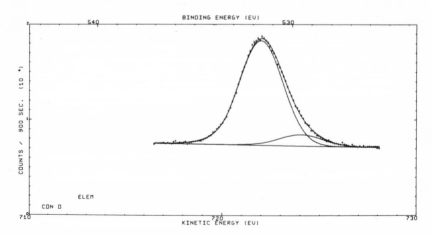

Figure 7. Oxygen 1s spectrum from untreated wool fiber.[55]

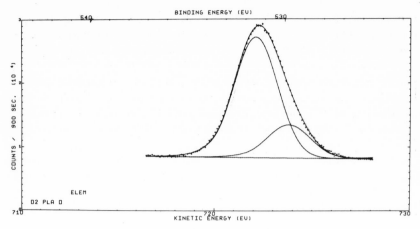

Figure 8. Oxygen 1s spectrum from O_2 plasma-treated wool.[55]

carbon compounds containing sulfur. The majority of these compounds contain oxygen bound to sulfur, carbon, and nitrogen. The O 1s line in compounds containing oxygen bound to sulfur and in carbonyl and carboxylate groups is around 532 eV. The high binding energy line around 532.2 eV is probably due to lines arising from oxygen in carbonyl, carboxylate, sulfone, sulfoxide, and sulfonic acid. This interpretation is supported by the presence of electron lines due to oxidized sulfur. The S 2p electron spectra from plasma-treated samples contain a strong line at 168 eV due to sulfonic acid or sulfate, indicating the presence of oxygen on sulfur. The lower binding energy line around 530.5 eV in Figure 8 is the region reported for metallic and refractory oxides. The O 1s electron line has been reported at 530.5, 530.7 eV for metal aluminates,[70] 530.7 eV for UO_2,[78,80] and 530.9 eV for chemadsorbed Pb.[81] Grunthaner[82] observed

Table 4. Oxygen 1s Electron Line Parameters, Spectra from Plasma-Treated Wool Fibers

Sample	Binding energy (eV)	Relative area	Shift in peak position (eV)
Untreated wool fiber	530.04	1.0	0.0
	532.06	9.19	2.02
Nitrogen plasma-treated wool fiber	530.73	1.0	0.0
	532.26	2.89	1.52
Afterglow plasma-treated wool fiber	530.87	1.0	0.0
	532.41	3.1	1.53
Oxygen plasma-treated wool fiber	530.61	1.0	0.0
	532.29	3.62	1.68

O 1s electron lines at 532.2 and 530.9 eV from oxidized iron dithiolene complexes. The latter line was attributed to oxide oxygen and the former to sulfoxide oxygens. Barber *et al.*[83] reported O 1s spectra from a graphite that had been exposed to atomic oxygen at room temperature and 673°K. Two O 1s lines were present. The higher binding energy line was attributed to carbonyl oxygen and the lower binding energy line to oxygen with a negative charge. The low binding energy 1s line increases in intensity by a factor of 3 upon plasma treatment. Apparently, a surface oxide species is formed on the surface of the wool fiber as a result of plasma treatment.

The N 1s electron line spectra are shown in Figures 9 and 10 and the parameters obtained from the deconvoluted spectra are given in Table 5. Two lines are present in these spectra, and the ratio of the two lines changes considerably upon plasma treatment. The high binding energy line decreases by a factor of three in intensity upon plasma treatment. This behavior is very similar to that observed for the O 1s line spectra.

A correlation of these lines with some structural features on the surface of the wool fiber is again difficult. The binding energy shifts for nitrogen are small, and a considerable number of structurally different nitrogen atoms could be present on the side chains of the amino acids on the surface of the fiber. In addition to the nitrogen atoms on the side chains, nitrogen in the polypeptide chain could also give rise to a nitrogen electron line. Few reports exist in the literature concerning the nature of the N 1s electron lines observed from proteins. Klein and Kramer[84] reported the N 1s spectra of the endoplasm from light-red kidney bean and L-isoleucyl L-alanine together with cystine. The nitrogen line for the dipeptide could be deconvoluted into two lines. A higher binding energy line coinciding in position with the line from cystine was assigned to the amino nitrogen. The line at lower binding energy was assigned to amide nitrogen formed during the peptide bonding. The nitrogen line from the kidney bean endoplasm was deconvoluted into two lines. A low-intensity line at lower binding

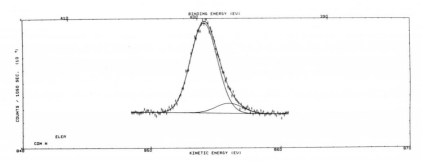

Figure 9. Nitrogen 1s spectrum from untreated wool fiber.[55]

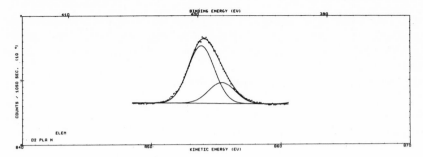

Figure 10. Nitrogen 1s spectrum from oxygen plasma-treated wool.[55]

energy was attributed to amide nitrogen, and the intense line at higher binding energy was attributed to amino nitrogen. Clark *et al.*[36,85] measured the N 1s electron line spectra of barley meals and polyamino acids.

One possible interpretation of the nitrogen 1s line spectra observed from wool fiber would be to attribute the lower line to the amide nitrogen of the polypeptide and to assign the higher line to the other nitrogen atoms in the side atoms. Upon plasma treatment some of the side-chain nitrogen-containing groups could be broken away by ion etching or oxidation leaving more exposed polypeptide nitrogen. This would explain the decrease in intensity of the high binding energy line upon plasma treatment.

Nordberg *et al.*[86] have reported N 1s electron shifts for a large number of compounds. The N 1s line for organic amines is in the region of 398.0 eV. Protonation of the amine shifts the N 1s line to 400.4 eV. The low line observed from the wool fiber could be due to amino-type nitrogen and the higher line due to protonated amine. Plasma treatment could reduce the number of protonated amines, thus leading to a decrease in the intensity at the higher binding line. These interpretations are ten-

Table 5. Nitrogen 1s Electron Line Parameters, Spectra from Plasma-Treated Wool Fiber

Sample	Binding energy (eV)	Relative area	Shift in peak position (eV)
Untreated wool fiber	398.03	1.0	0.0
	399.94	8.9	1.9
Nitrogen plasma-treated wool fiber	398.6	1.0	0.0
	400.3	3	1.67
Oxygen plasma-treated wool fiber	398.95	1.0	0.0
	400.54	2.74	1.59

tative and much more work will have to be done in order to understand the nature of the nitrogen electron spectra from proteins.

6.5. Surface Composition from Electron Line Intensities

The electron line intensities can be related to atomic composition using Wagner's sensitivity factor.[25] The carbon, oxygen, and nitrogen atomic ratios on the surface can be calculated from the relative elemental line intensities; these data are given in Table 6. The elemental composition of wool fiber obtained by chemical analysis is $C_1O_{0.33}N_{0.28}$. This ratio reflects the composition of the bulk. The elemental ratio for the control calculated from electron line intensities is $C_1O_{0.33}N_{0.12}$. This ratio for oxygen-plasma-treated fiber is $C_1O_{0.96}N_{0.53}$, and for nitrogen-plasma-treated fiber the ratio is $C_1O_{0.55}N_{0.36}$. The elemental ratio calculated from XPS surface analysis is in fair agreement with that calculated from bulk analysis. The nitrogen content is low. The surface elemental ratio changes appreciably upon plasma treatment. Approximately three times as much oxygen and four times as much nitrogen are on the surface after oxygen plasma treatment. The increase in the ratio of oxygen to carbon correlates with the increase in the carbon electron lines at higher binding energy.

Table 6. Elemental Electron Line Intensities

Sample	Maximum line intensity (cps)	Total line intensity	Corrected line intensity
Control			
Carbon	48,600		180,000[a]
Oxygen	31,100		59,700[b]
Nitrogen	8,780		21,000[c]
Oxygen plasma			
Carbon	23,300	39,100	145,000[a]
Oxygen	56,700	72,300	139,000[b]
Nitrogen	24,100	32,400	77,000[c]
Nitrogen plasma			
Carbon	31,500	50,000	185,000[a]
Oxygen	39,700	52,500	100,000[b]
Nitrogen	21,400	28,200	67,000[c]
Afterglow			
Carbon	28,200	44,300	164,000[a]
Oxygen	51,250	68,000	131,000[b]
Nitrogen	14,250		

[a] Intensity divided by 0.27.
[b] Intensity divided by 0.52.
[c] Intensity divided by 0.42.

6.6. Plasma Deposition of Fluorocarbon Coatings

Coatings can improve the properties of fibers, and we have utilized plasma polymerization of fluorocarbon monomers to produce fluorocarbon coatings on fibers. The polymerization of organic monomers in plasmas has been reviewed,[87] as well as the use of fluorocarbon films to impart oil and water repellency and surface and soil resistance to fabrics.[88,89]

XPS is particularly suited for the analysis of fluorocarbon coatings. The fluorine atom has a high photoelectric cross section,[25] and the fluorine atom shifts the binding energy of carbon so that carbon atoms containing one, two, or three fluorine atoms can be differentiated.[90]

Data are presented from plasma-polymerized fluorocarbon films on polystyrene, polypropylene, and wool textiles. The plastic substrates represent the case of a more smooth and inert surface, while wool textile is an example of an extremely uneven surface containing various reactive functional groups. It has been generally observed that core-electron line shapes are related to the smoothness and regularity of the surface.[91] As anticipated, electron line widths were found to be more uniform from films deposited on polystyrene and polypropylene. The experimental details of the plasma polymerization and deposition of hexafluoroethane and tetrafluoroethylene are given in the original paper.[92]

XPS spectra were measured on a Varian IEE 15 spectrometer with a MgK_α X-ray source. Spectra were deconvoluted using the least-squares computer program mentioned earlier.[29,30] Carbon and oxygen spectra were fitted with Lorentzian functions with a leading constant tail. Fluorine spectra were fitted using Gaussian functions with a leading constant tail. Unless otherwise specified, line widths were fixed at 2.5 eV.

6.7. Spectra from Fluorocarbon Films on Polypropylene and Polystyrene

6.7.1. Carbon 1s Spectra

Typical C 1s electron spectra obtained from plasma-polymerized hexafluoroethane on polymer substrate are presented in Figure 11. The data calculated from the spectra such as line position, half width, and relative intensities are given in Table 7. The spectra were fitted using five lines. The rationale for a five-line model is the following. The binding energies of carbon in fluorocarbon polymers containing CF_3, CF_2R, and CFR_2 groups, where R is hydrogen or alkyl, have been reported. The C 1s binding energy in the CF_3 group is around 294 eV and the CF_2 carbon is 292 eV in tetrafluoroethylene.[80] Polymers containing CFR_2 groups yield binding energies between 289.4 and 292.8 eV, depending on the electro-

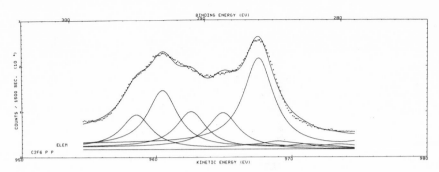

Figure 11. Carbon 1s spectrum from plasma polymerized hexafluoroethane deposited on polyethylene.[72]

negativity of the R group.[90,93] The line at lowest binding energy is assumed to arise from aliphatic carbon and is used as a reference at 285.0. There are several functional groups containing carbon that give rise to binding energies between 285 and 290 eV. Carbon containing a single bond to oxygen, such as alcohols or ethers, occurs around 286.5–286.8 eV.[76] Carbon in ketones is around 288.1 eV, while carbonyl carbon in acids or esters occurs around 289.5 eV.[76] The binding energy of carbon is shifted to higher energies when fluorocarbon groups are attached, and carbon adjacent to carbon bonded to fluorine occurs in this range.[90]

The spectra were fitted with a fifth line whose position ranged between 287.2 and 287.8 eV. There is little doubt that the carbon electron lines in the region of 294 and 292 eV are due to CF_3 and CF_2R carbons. The carbon electron line around 290 eV is probably a combination of lines arising from carbonyl carbon and carbon containing one fluorine. The line in the region of 287.5 eV is probably a composite of lines due to carbon singly bonded to oxygen and carbon attached to carbon bonded to fluorine atoms. The positions of the two lines above 285.0 eV are probably least certain; however, a reasonably consistent fit of the spectra was obtained. Linewidths ranged from 2.43 to 2.69 eV for the fluorocarbon films deposited onto polymer substrates.

6.7.2. Oxygen 1s Spectra

Polymers formed in plasmas are readily oxidized on the surface after exposure of the film to the atmosphere.[87] We have previously reported that fluorocarbon polymers obtained by plasma polymerization of perfluorobutene-2 contain a high concentration of unpaired spins, and these radicals probably react with oxygen yielding oxygen-containing species on the surface.[74] The assignment of carbon electron lines in the region between 286 and 290 eV to carbon functional groups containing oxygen is also supported by the detection of oxygen on the surface of these films.

An example of the O 1s electron spectra obtained from the fluoro-carbon films on polymer substrate is given in Figure 12. Oxygen 1s electron line parameters obtained from the spectra are presented in Table 7. The O 1s electron line spectra were deconvoluted into two lines for all spectra other than the spectra obtained from tetrafluoroethylene films on polymer substrate. The O 1s electron spectra consisted of two lines at 532.0 and 533.4 eV for C_2F_4 polymer on polystyrene and 532.4 and 534.5 eV for C_2F_6 polymer on polypropylene. The O 1s electron binding energy in most organic functional groups occurs around 532 eV.[78] This includes oxygen in alcohols, ethers, ketones, and carboxylic acids. The O 1s line at higher binding energy was only observed on films containing fluorine. We suggest that the higher binding energy O 1s line arises when oxygen is bonded to carbon atoms containing fluorine such as CF_3—O—CF_2—. The electron-withdrawing fluorocarbon groups could increase the binding energy of the oxygen atom.

6.7.3. Fluorine 1s Spectra

The F 1s electron line appeared to be symmetrical for the fluorocarbon films on polypropylene and polystyrene. The F 1s electron line from fluorocarbon films polymerized from hexafluoroethane had a width of 2.75 eV. The width of the fluorine lines obtained from films from poly-merized tetrafluoroethylene ranged in width from 3.0 to 3.3 eV. In general,

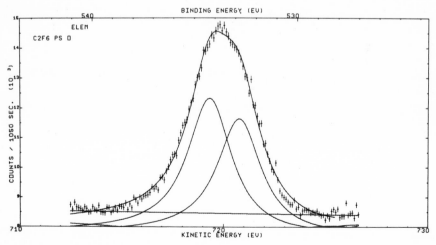

Figure 12. Oxygen 1s spectrum from fluorocarbon film deposited on polystyrene.[72]

Table 7. Electron Line Parameters from the Spectra of Plasma Polymerized Films

Sample	C 1s intensity (10³ cps)	Carbon 1s binding energy (eV) (relative line ratio)					FWHM (eV)	O 1s intensity (10³ cps)
1. C_2F_6 on polypropylene	8.0	285.0 (1)	287.6 (0.402)	290.0 (0.415)	292.1 (0.648)	294.0 (0.380)	2.69	2.373
2. C_2F_6 on polypropylene	7.82	285.0 (1)	287.6 (0.426)	289.8 (0.336)	292.1 (0.508)	294.1 (0.316)	2.72	3.730
3. C_2F_6 on wool 31C	4.779	285.0 (1)	286.0 (0.37)	288.3 (0.22)	290.1 (0.105)	292.9 (0.124)	2.26	7.230
4. C_2F_6 on wool 31B	4.124	285.0 (1)	286.4 (2.5)	287.7 (0.0008)	289.0 (0.505)	290.4 (0.008)	1.7	4.467
5. C_2F_6 on wool 31E		285.0 (1)	285.9 (1.28)	287.7 (2.7)	289.7 (4.7)	291.7 (0.0005)	2.12	
6. C_2F_4 on polypropylene	18.231	285.0 (1)	287.2 (2.2)	289.3 (1.42)	291.6 (2.62)	293.5 (1.41)	2.66	4.905
7. C_2F_4 polypropylene, no argon, D	14.7	285.0 (1)	287.8 (0.58)	289.8 (0.37)	292.3 (0.61)	294.3 (0.43)	2.43	6.450
8. C_2F_4 on wool 30B	2.06	285.0 (1)	286.8 (0.85)	289.7 (0.40)	291.0 (0.6)	292.9 (0.6)	3.19	1.580
9. C_2F_4 on wool 30E	2.21	285.0 (1)	286.6 (0.8)	288.6 (0.40)	291.1 (0.57)	293.1 (0.46)	3.27	1.432

Sample	O 1s B.E. eV (ratio)	FWHM (eV)	Total intensity (10³ cps)	Corrected intensity (10³ cps)	F 1s intensity (10³ cps)	F 1s B.E. eV (ratio)	FWHM (eV)	Surface atom ratio
1. C_2F_6 on polypropylene	532.4 (1) / 534.5 (0.54)	3.49	3.660	7.000	116.5	689.2	2.8	16.6 F/O
2. C_2F_6 on polypropylene	532.0 (1) / 533.4 (1.2)	2.6	6.830	13.600	96.3	689.3	2.67	7 F/O
3. C_2F_6 on wool 31C					27.2	686.5 (1) / 688.9 (3.9)	2.6	2 F/O
4. C_2F_6 on wool 31B	532.7 (1) / 533.9 (0.62)	2.6	7.240	13.900	4.95	687.1 (1) / 689.5 (0.5)	2.6	0.35 F/O
5. C_2F_6 on wool 31E	532.6 (1) / 534.2 (0.4)	2.6			3.40	687.1 (1) / 689.0 (1.7)	2.6	
6. C_2F_4 on polypropylene					160.0	688.8	3.3	
7. C_2F_4 polypropylene, no argon, D					97.8	689.3	3.3	
8. C_2F_4 on wool 30B	532.4 (1) / 534.2 (0.8)	3.19	2.820	5.420	24.1	688.3	3.3	4.45 F/O
9. C_2F_4 on wool 30E	531.6 (1) / 533.3 (0.5)	3.27	2.219	4.270	22.6	687.7	3.0	5.3 F/O

the fluorine electron line was quite intense from these films and ranged in intensity from 96,000 to 160,000 counts/sec.

6.8. Spectra from Fluorocarbon Films on Wool Textile

6.8.1. Carbon 1s Spectra

Using plasma polymerization conditions similar to those employed to polymerize films on polymer substrate, the fluorocarbon film thickness and density of coverage were much lower when the substrate was wool textile. The low level of fluorocarbon film coverage was indicated by much lower fluorine electron line intensities and weaker carbon electron lines at higher binding energies. The carbon electron lines were shifted about 1 eV lower than measured on fluorocarbon films on polymer substrate. The binding energy of carbon bonded to fluorine not adjacent to other fluorocarbon groups is lower than when the adjacent carbon is also fluorinated. Clark et al.[90] have studied carbon binding energies in fluorocarbon polymers. For example, the binding energy of CF_2 carbon adjacent to CH_2 is 290.8 eV, while carbon in tetrafluoroethylene is 292.3 eV.

Table 7 contains the results of the deconvolution of three carbon 1s electron spectra from wool fiber exposed to a hexafluoroethane plasma. The intensity of the higher binding energy carbon lines as well as the fluorine electron line intensity varied considerably for these samples. Sample 31C, whose carbon spectrum is shown in Figure 13 had a higher fluorine 1s line intensity and contained more intense higher binding energy carbon lines. Electron spectra from samples 31B and 31E consisted of weak fluorine lines and low-intensity higher binding energy carbon lines. The five lines from spectra 31C are interpreted as follows. In addition to the reference hydrocarbon line at 285 eV, the line at 286.0 eV could contain

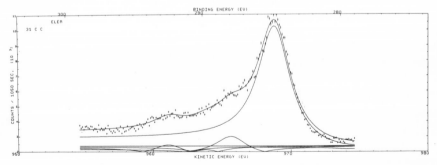

Figure 13. Carbon 1s spectrum from fluorocarbon film deposited on wool fiber.[72]

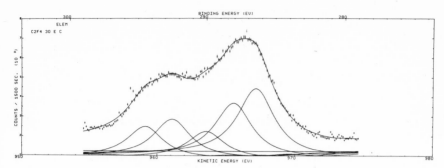

Figure 14. Carbon 1s spectrum from fluorocarbon film deposited on wool fiber (sample 30E).[72]

varying contributions from carbon containing single-bonded oxygen or non-fluorine-containing carbon adjacent to carbon substituted with fluorine. The next three higher binding energy lines arise from carbon atoms bonded to one, two, or three fluorine atoms not adjacent to their fluorine-containing carbon atoms. The carbon line at 288.3 eV could contain some contribution from carbon containing a double bond to oxygen. The deconvoluted spectrum from sample 31B contains a single higher binding energy carbon line at 286 eV in the binding energy region expected for carbon bonded to one fluorine atom. This sample probably consisted of randomly distributed carbon bonded to one fluorine atom. The carbon electron spectra from wool fiber exposed to a tetrafluoroethylene fluorocarbon plasma are interpreted in the same way as the carbon spectrum from sample 31C. The carbon electron spectrum from sample 31E is given in Figure 14.

6.8.2. Oxygen 1s Spectra

Oxygen 1s lines measured on the wool textiles exposed to the fluorocarbon plasmas are given in Table 7, and a typical spectrum is shown in Figure 15. These lines are believed to arise from the same functional groups discussed in the previous section on the interpretation of the oxygen spectra obtained from the plasma-polymerized fluorocarbon films deposited on polymer substrate.

6.8.3. Fluorine 1s Spectra

The F 1s spectra obtained from the fluorocarbon-coated wool fibers are also included in Table 7. The F 1s electron spectra obtained from plasma-polymerized hexafluoroethylene polymer on the wool fiber substrate contained shoulders, and the spectra were deconvoluted into two

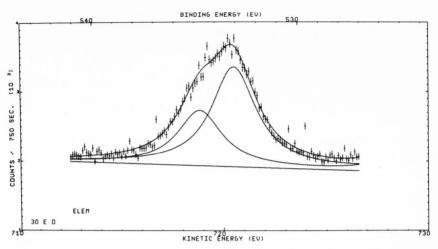

Figure 15. Oxygen 1s spectrum from fluorocarbon film deposited on wool fiber (sample 30E).[72]

lines. The deconvoluted F 1s spectrum from sample 31E is shown in Figure 16. The high binding energy 1s line around 689 eV is in the region reported for fluorine covalently bound to carbon.[90–94] The lower binding energy line around 687 eV is in the region reported for fluoride ion.[94] The fluorine electron lines measured on wool textile exposed to tetrafluoroethylene plasma were somewhat broader than those obtained from the other fluorocarbon surfaces, but symmetrical in shape. These lines were not deconvoluted.

6.9. Surface Composition from Electron Line Intensities

Some idea of the surface composition of the fluorocarbon films can be obtained from the intensities of the fluorine and O 1s lines. The

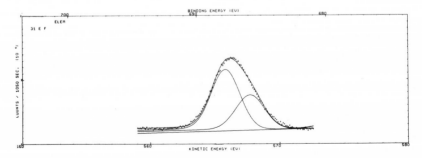

Figure 16. Fluorine 1s spectrum from film deposited on wool fiber (sample 31E).[72]

calculated fluorine-to-oxygen atom ratio for several plasma-polymerized fluorocarbon surfaces is given in Table 7. The electron line intensities were divided by Wagner's atomic line intensity correction factors.[25] This procedure usually yields atomic ratios to within 5%. The Fluorine-to-oxygen atom ratio on the surface ranged from 16.6 to 0.3, indicating considerable variation in the fluorocarbon character of the surface as well as the amount of surface oxidation.

6.10. Conclusions

Combining the surface core electron line intensity data and the relative electron line positions, a fairly detailed picture of the surface resulting after exposure to fluorocarbon plasmas can be obtained. Plasma polymerization of hexafluoroethane and tetrafluoroethylene onto polymer substrate yielded fluorocarbon surfaces with the following characteristics. The fluorine electron lines were quite intense, and fairly intense high binding energy carbon electron lines associated with carbon bonded to fluorine were present. The carbon binding energies were similar to those obtained from fluorocarbon surfaces containing adjacent fluorinated carbon atoms. Carbon atoms containing one, two, and three fluorine atoms are present on the surface, indicating considerable monomer breakdown under the plasma conditions employed here. Oxygen was detected on the surface, providing further evidence for surface oxidation of plasma-polymerized films.

The surface of wool fibers exposed to these fluorocarbon plasmas were quite different. The fluorine electron lines were generally much less intense. The carbon electron lines in the region of carbon bound to fluorine were shifted to lower binding energies, indicating that the fluorinated carbon atoms were not bonded to other carbon atoms containing fluorine. Carbon atoms containing fluorine are probably randomly distributed among nonfluorinated carbons. The extent of fluorocarbon film coverage was much less in this case.

7. Detection of Chemically Modified Proteins

Certain chemical modifications of proteins can be readily detected by XPS. If the chemical modification contains an element such as iodine not present in the native protein or an element whose binding energy is shifted from the native protein, then a new core-electron line appears in the spectrum that is proportional to the extent of chemical modification.

Core-electron line intensities can be related to atom ratios using experimentally determined elemental sensitivity factors. Although the relationship between line intensity from surfaces and elemental composition is far from straightforward, the elemental ratios calculated from XPS line intensities for pure proteins are in good agreement with theoretical ratios.[47] One must assume that the surface composition is representative of the entire sample. Two examples of the detection of chemical modifications on proteins will be reviewed.

Fluorine has a relatively high atomic photoelectron cross section,[25] and chemical modifications containing fluorine give rise to intense photoelectric lines. The ϵ-amino group of lysine can be specifically modified using ethyl thiol trifluoroacetate, and the lysine groups present in bovine serum albumin (BSA) were acetylated according to this procedure.[95] The F 1s electron line from trifluoroacetylated BSA is shown in Figure 17. The F 1s line is 5800 counts above background from a single scan and is more than sufficient for an accurate line intensity measurement. The nitrogen-to-fluorine atom ratio calculated from XPS line intensity data and the nitrogen-to-fluorine atom ratio obtained from wet chemical analysis are shown in Table 8.

The average values of the elemental ratios as determined by XPS data and wet chemical analysis agree within 5%.

Ethyl vinyl sulfone-modified BSA has also been studied by XPS. The sulfone group contains sulfur with a binding energy that is higher than

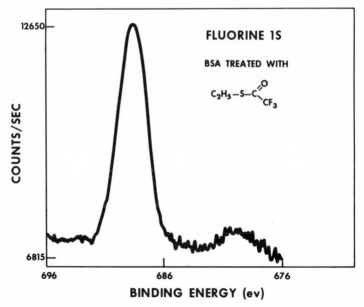

Figure 17. Fluorine 1s spectrum from trifluoroacetylated bovine serum albumin.[79]

Table 8. Nitrogen and Fluorine Core Electron Line Intensity and Wet Chemical
Analysis Data for N-Trifluoroacetylated Bovine Serum Albumin

Sample	N 1s (cps)	Corrected intensity[a] (cps)	F 1s (cps)	N/F atom ratio	N/F atom ratio, wet chemical analysis
M-196					
Run 1	15,800 ± 100	37,600	9,700 ± 200	3.88	4.45
Run 2	20,800 ± 100	49,600	10,700 ± 100	4.63	
M-197					
Run 1	12,900 ± 100	30,700	6,450 ± 100	4.75	
Run 2	16,700 ± 100	39,750	8,350 ± 100	4.76	5.02

[a] Observed intensity divided by Wagner's sensitivity factor.[25]

the binding energy of sulfur in BSA, and a distinct line appears in the
sulfur region due to the sulfone sulfur.

The S 2p spectra from BSA and ethyl vinyl sulfone-modified BSA are
shown in Figure 18. The higher binding energy line is related to the extent
of chemical modification, and the agreement between wet chemical analysis

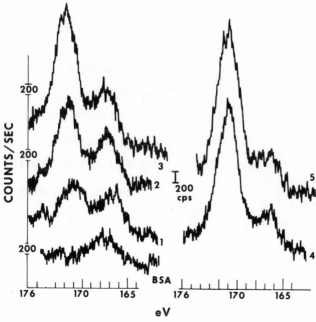

Figure 18. Sulfur 2p spectra of native BSA and BSH modified with increasing amounts
of ethyl vinyl sulfone.[95]

and calculations from XPS line intensities was quite good.[96] The intensity of the sulfone sulfur line in chemically modified BSA was quite sensitive to exposure of the sample to radiation in the instrument.

8. Bacterial Cells and Envelopes

8.1. Structure of Envelopes of Gram-Positive and Gram-Negative Bacteria

Bacterial cell walls or envelopes represent an interesting area for the application of XPS for surface analysis. The two major classifications of bacteria based on the Gram stain procedure, namely gram-positive and gram-negative, differ considerably in the structure of their outer envelope.[97] Gram-positive bacteria contain a single membrane with an exterior of peptidoglycan containing teichoic acids and lipoteichoic acid. The outer surface of a typical gram-positive bacteria is shown schematically in Figure 19. A representative peptidoglycan is shown in Figure 20. Teichoic acids are phosphorus-containing polysaccharides with some amino acid content, and these are shown schematically in Figure 21. There is some uncertainty about the distribution of teichoic acid in the peptidoglycan in the cell wall.[98]

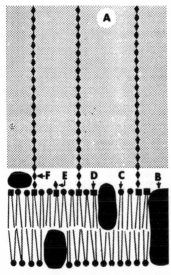

MEMBRANE COMPONENTS:

B = PROTEIN

C = PHOSPHOLIPID

D = GLYCOLIPID

E = PHOSPHATIDYL GLYCOLIPID

F = LIPOTEICHOIC ACID

Figure 19. Diagrammatic representation of the cell wall (A) and membrane of a gram-positive organism.[111]

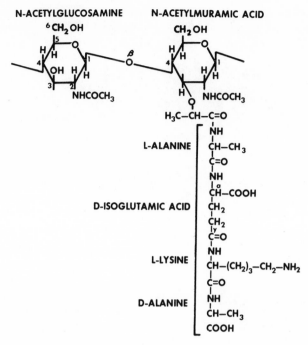

Figure 20. Repeating unit of a bacterial cell wall peptidoglycan.[112]

Gram-negative organisms have a much more complex outer structure. There are two lipid bilayer membranes with lipopolysaccharides in the outer surface and peptidoglycan between the inner and outer membrane. A schematic diagram of the outer layer of a typical gram-negative organism is shown in Figure 22.[99] Lipopolysaccharides have a complex structure, and a schematic representation is given in Figure 23.[100]

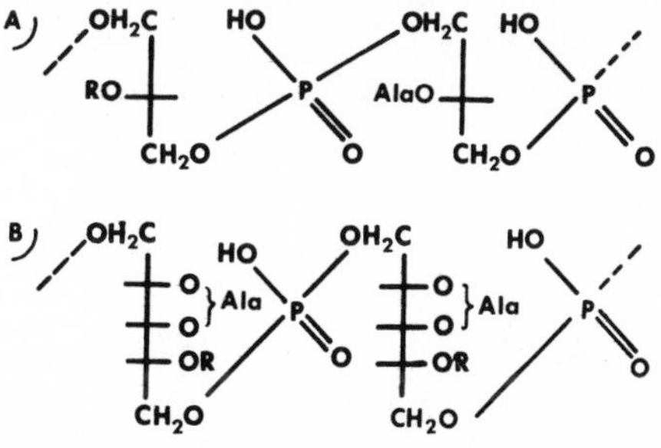

Figure 21. Repeating units of bacterial cell wall teichoic acids.[14]

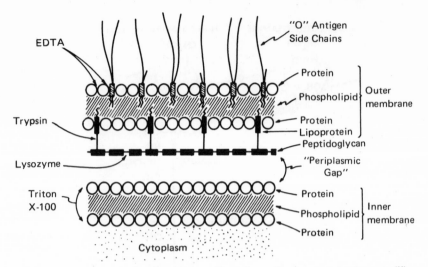

Figure 22. Schematic diagram of the cell envelope of gram-negative bacteria.[99]

8.2. Electron Spectra and Surface Composition of Selected Bacteria

Four species of bacteria were chosen for study: *Bacillus subtilis* 168, (BS); *Bacillus megaterium* KM (BM); *Micrococcus lysodeikticus* (ML); and *Escherichia coli* B (*E. coli*).[101] The first two are gram-positive species while ML contains no teichoic acid and *E. coli* is a gram-negative bacteria. Washed suspensions of bacterial cells were deposited on the support disk and air dried. Substrates consisted of polystyrene polybutadiene plastic, Resin K (Phillips Petroleum), or glass disks cut from microscope coverslips and coated with SiO in a vacuum evaporator. Substrate coverage could be estimated from the intensity of the silicon lines from the substrate. A typical survey scan of *B. megaterium* KM cells on SiO/glass disk is shown in Figure 24. The electron lines detected and their relative intensity for the four species of bacteria studied are given in Table 9. The core-elemental line intensities were converted to atom ratios using Wagner's sensitivity factors. The surface atom ratios varied considerably for the four species. The intensity of the surface phosphorus line was greatest for BS and BM and least for *E. coli*. The individual surface phosphorus electron lines are shown in Figure 25 A Sodium was the common cation found on the cell surfaces. Potassium was found only in small amounts. Isolated BS cell walls showed a very strong sodium signal.

8.3. Oxygen Plasma Etching and Elemental Depth Profile Curves

To investigate the change in concentration of the various elements as a function of depth into the cell wall surface, the four species of bacteria

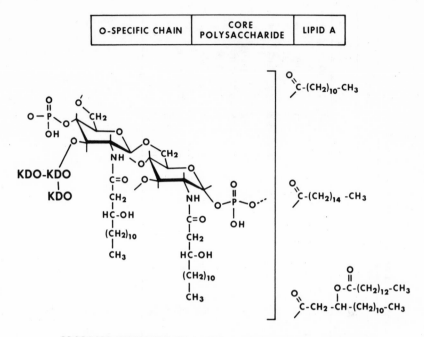

PROPOSED STRUCTURE OF A LIPID-A UNIT OF THE S. MINNESOTA R595 GLYCOLIPID WITH AN ATTACHED KDO TRISACCHARIDE.

Figure 23. Schematic representation of a bacterial lipopolysaccharide.[100]

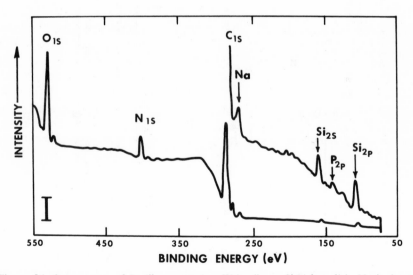

Figure 24. Survey scan of *Bacillus megaterium* KM cells on SiO/glass disk. Na is Auger KLL line of Na. Bar = 1000 cps (lower scan) or 125 cps (upper).[36]

Table 9. Line Intensity and Atom Ratio of Surface Elements of Bacteria as Obtained by ESCA[a]

Specimen	Specimen support	Disk coverage (%)	Line intensity[b] (10^3 cps)						Atom ratios[e]				
			Si_{2p}	C_{1s}[c]	O_{1s}	N_{1s}	P_{2p}	Au_{4f7}[d]	N	C	O	P	Si
None	Resin K	0	0	41.6	2	0	0	100	0	13.4	1	0	0
None	SiO/Glass	0	25	25.2	165	0	0	190	0	0.9	2.2	0	1
Growth medium	SiO/glass	100						125	1[f]	6.5[f]	8.7[f]	0.2[f]	0[f]
Cells													
E. coli B	SiO/glass	78.8	5.4	64.5	79	13.9	1	190	1	9.3	2.7	0.07	
E. coli B	Resin K			24.6				100	1[f]		2.3[f]		
M. lysodeikticus	SiO/glass	96.3	0.9	36	63.5	15	0.2	190	1	6.4	2.0	0.005	
B. megaterium KM	SiO/glass	93.0	1.6	56	57	11.5	0.6	190	1	13.3	3.7	0.05	
B. megaterium KM	Resin K		0		20.3	4.8	0.4	100	1		3.6	0.08	
B. subtilis 168	Resin K		0	24	14.9	3.1	0.6	100	1		3.7	0.21	
Cell walls													
B. subtilis 168	SiO/glass	94.4	2.4	52	83.5	7.2	3.2	190	1	17.5	7.5	0.41	
N-acetylglucosamine	Bulk specimen[g]			8.8	19.6	2.3			1[h]	9.6[h]	6.6[h]		
Chitin	Bulk specimen			24	38.8	5.6			1[h]	11.2[h]	5.1[h]		

[a] Reproduced from reference 36.
[b] Single element scan, not corrected for C contamination or disk coverage.
[c] C_{1s} peak height measured at 284 eV.
[d] External (machine) gold standard.
[e] Ratios based upon peak areas, corrected for disk coverage, except [e].
[f] Ratios based on peak heights estimated from multiple scan spectrum.
[g] Compressed pellet.
[h] Not corrected for C contamination.

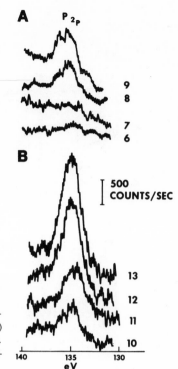

Figure 25. (A) P 2p line of *E. coli* cells (6), *M. lyso-deikticus* (7), *B. megaterium* (8), and *B. subtilis* (9) cells.[85] (B) 10–13: Etch series of P 2p line of *B. megaterium* cells (sequential etching of the same specimen for 0, 2, 4, and 6 min).[101]

were subjected to OPE and survey scans as well as single scans over various elements recorded. A series of survey scans as a function of OPE exposure time for *E. coli* is shown in Figure 26. Etching added the elemental lines of P, S, K, Na, and Ca to that of CNO found on the surface. A series of single-scan phosphorus lines is given in Figure 25b for BM as a function of OPE time. Before these data can be converted into a depth profile, the rate of etching and the uniformity of etching must be known. The individual rate of etching is difficult to measure for each type of substrate; however, the rate of etching was calibrated by adding polystyrene spheres to the surface of the specimen and the decrease in diameter of the spheres measured as a function of etching time. This procedure is discussed in Section 4. The uniformity of etching was investigated by studying electron micrographs of bacteria specimens taken after various etching times.[45] We concluded that the etching process can become nonuniform after 5–10 min etching in many cases. Electron micrographs of *B. subtilis* cells before and after 5 min of OPE are shown in Figure 27. As a result, etching times were usually restricted to slightly more than 5 min. Subject to these restrictions, depth profile curves are shown in Figure 28 for those bacteria for

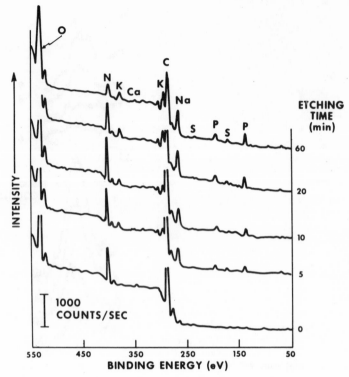

Figure 26. Survey scans of *E. coli* B cells on resin K support disk after different etching times.[101]

several elements. Electron microscopic analysis of the etched cells or cell walls indicated that after 30 sec etching the isolated walls appeared thinner, but both they and the intact cells still showed a smooth surface. Etching for 1 min gave a pitted aspect to the isolated walls. Isolated cell walls disappeared completely after 2 min etching. Etching for 5 min or longer clearly removed the walls from intact cells and selectively attacked the interior structures. The phosphorus signal exhibited a plateau for the four species. There was no feature of the curve for *E. coli* to suggest that two membranes were present in this species. At present the interpretation of these profile curves is quite tentative, and model studies on lipid bilayers may be of some use in investigating the usefulness of this approach.

The distribution of calcium in BM spores was investigated, and a wide-scan spectrum from the surface and after etching for 5 min is shown in Figure 29. Calcium is absent in the surface scan and fairly intense after 5 min etching.

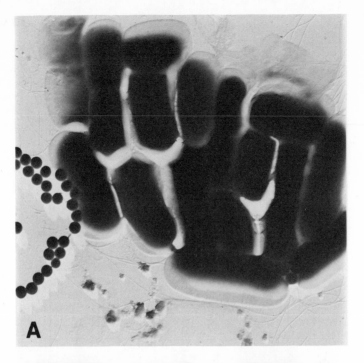

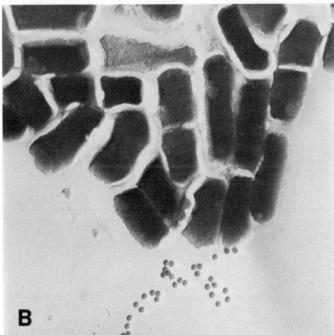

Figure 27. *B. subtilis* cells with pSL: (A) unetched, (B) etched 5 min. Both shadowed.[104,46]

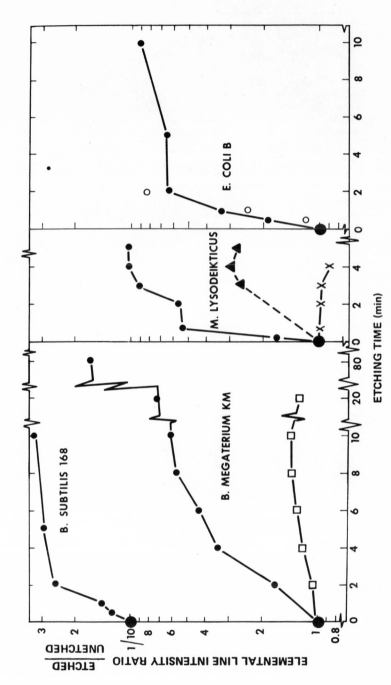

Figure 28. Influence of the sequential etching time on the signal intensity of elements. Specimen supports: Resin K (*B. subtilis*), SiO/glass (others). ●——● P_{2p}, □——□ Si_{2p}, ▲- -▲ K_{2p}, X—X N_{1s}, ○ same as ● but single etching of different specimens.[101]

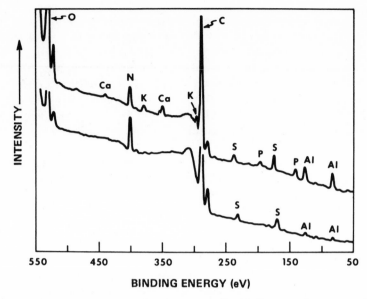

Figure 29. XPS spectra of 32-mm² surface covered with *B. megaterium* spores. Lower curve; unetched. Upper; same surface etched 5 min.[46]

9. Tissue-Culture Cells

The importance of the surface of cells has been recognized in such currently important fields as cancer research[5,7,12] and immunology.[10,11] The fact that the composition of cell-surface components such as antigens and carbohydrates changes as a result of viral transformation has focused attention on the nature of the components on the surface of cells, techniques for identifying and studying cell-surface components, and the role of these cell-surface components in cell recognition, cell movement, and cell division.[6]

Cell-surface-labeling techniques have been developed to label and visualize specific components on the surface of cells using scanning electron microscopy (SEM).[102] Markers such as ferritin, hemocyanin, and tobacco mosaic virus can be coupled to molecules that bind specifically to certain receptors on the surface such as concanavalin A or antibody. Electron probe microanalysis has also been used to analyze cells and tissues.[33]

9.1. Specimen and Substrate Considerations

Tissue-culture cells were chosen for study since the surface properties of cells are of considerable importance and these cells have the desirable property of growing on various substrates as monolayers.[37] It was also of interest to see if changes in the surface composition of cells before and after viral transformation could be detected by XPS. The Balb 3T3 A31 HYF cells will subsequently be referred to as normal cells. The MSV/MLV Balb 3T3 A31 HYF cells will subsequently be referred to as transformed cells. Cells were grown on polystyrene disks 1/4 in. in diameter cut from tissue culture dishes or 1/4-in. glass disks cut from microscope coverslips. These disks could be readily mounted in the instrument with minimum specimen handling.

Different substrates were used to overcome the difficulties resulting from interference of elemental line intensity contribution from the substrate and the specimen covering the substrate. The spectrometer has a variable carbon contamination which always seemed to contribute a signal at 285 eV. Due to the uncertainty involved in assigning line intensities from the carbon line at 285 to the specimen, this line was not used to characterize the specimen surface.

Glass and polystyrene were used as substrates. Polystyrene has the disadvantage of contributing a substrate carbon line, while glass contributes an interfering line due to oxygen. The silicon line from glass is unique to the substrate and was used to estimate the degree of surface coverage by cells. The difference in the silicon intensity from clean glass disks and disks covered with cells was taken as a measure of surface coverage. The contribution of the oxygen signal from the glass substrate was estimated by measuring the oxygen-to-silicon line-intensity ratio for clean glass and assuming that the silicon intensity from the substrate containing cells had associated with it a similar oxygen intensity. Polystyrene has a negligible oxygen signal, and the oxygen intensity characteristic of cells could be obtained using polysytrene as the substrate. The oxygen signal from cells on polystyrene and from cells on glass with the appropriate substrate correction were in reasonable agreement.

The disruptive effects of dehydration and evacuation on biological samples are well documented in the electron microscopy literature.[104-107] After some preliminary studies on cells introduced directly into the spectrometer and evacuated, we adopted the standard procedures used to fix and critical-point-dry speciments prior to introduction into the spectrometer.

A description of the specimens studied and their treatment prior to introduction into the instrument is given in Table 10.

9.2. Electron Spectra from Unfixed Cells

In order to ascertain the sensitivity of electron line signals to the external composition of the cell surface, wide-scan spectra were obtained

Table 10. Description of Balb 3T3 Cell Specimen Preparations[a]

Specimen	Preparation	Substrate	Angle between specimen plane and detector
1 Normal cells	Growth medium decanted	Polystyrene	90°
2 Normal cells	Washed with isotonic saline solution	Polystyrene	90°
3 Normal cells	Washed with distilled water	Polystyrene	90°
4 Normal cells	Not fixed	Polystyrene	90°
5 Transformed cells	Not fixed	Polystyrene	90°
6 Normal cells	Fixed with glutaraldehyde ethanol critically dried with Freon 113	Glass mounted on gold holder	90°
7 Normal cells	Fixed with glutaraldehyde ethanol critically dried with Freon 113	Glass on metal mount	90°
8 Transformed cells	Fixed with glutaraldehyde ethanol critically dried with Freon 113	Glass on metal mount	90°
9 Normal cells	Fixed with glutaraldehyde ethanol critically dried with Freon 113	Glass	90°
10 Normal cells	Fixed with glutaraldehyde ethanol critically dried with Freon 113	Glass	15°
11 Transformed cells	Fixed with glutaraldehyde ethanol critically dried with Freon 113	Glass	90°
12 Transformed	Fixed with glutaraldehyde ethanol critically dried with Freon 113	Glass	15°

[a] Reproduced from reference 37.

over the region 550 eV to 50 eV for normal cells after several clean-up procedures. The position, intensity, and the structural interpretation for the photoelectron lines obtained from these specimens are given in Table 11.

In each of these spectra, the most intense lines are the O 1s at 532 eV, the N 1s at 400 eV, and the C 1s at 285 eV. Low-intensity lines due to chloride ion, 198.1 eV; sulfate ion, 168.5 eV; organic disulfied, 163 eV; phosphate, 133.2 eV; and sodium ion, 63 eV (Auger KLL 122 270 eV), are present in the spectra obtained from the unwashed cells. In the spectra from the cells washed with isotonic sodium chloride, the line due to chloride

Merle M. Millard

Table 11. Parameters Derived from Electron Spectra of Cells[a]

Sample[b]	O 1s	N 1s	C 1s	Cl 2p (Cl−)	S 2p (SO4−2)	S 2p (S-S)	P 2p (PO4−2)	Na 2s (Na+)	Oxygen/Nitrogen
1	532.2 (32.0)	400.0 (7.8)	285.8 (82.5)	198.1 (1.2)	168.5 (0.4)	162.5 (0.4)	133.2 (1.2)	63 (0.7)	2.78
2	531.5 (18.2)	400.0 (7.5)	285.0 (102.5)	198.4 (2.0)	Trace	163 (0.6)	134 (0.3)		1.98
3	532.2 (26)	400.2 (7.5)	285.0 (75)		Trace	Trace	133 (0.5)		2.78
4	(22.2)	(9)	(70)						2.01
5	(17.5)	(7.25)	(65)						1.99
6c	533 (24)	400.8 (5)	285 (70)			Trace			1.08

[a] Electron line positions are in units of electron volts. Intensities are in parentheses in units of 1000 cps.
[b] Numbers refer to specimens of Table 10.
[c] Additional lines observed in the spectra were Sn 2p3 (496) Sn 3s (488) Pb 4f 5/2, 7/2 (144, 138) Si 2p (103) Au 4f 7/2, 5/2 (85, 83.0).

ion increases in intensity while the electron lines due to the sulfate and phosphate decrease in intensity. All lines due to those ions except phosphate are essentially absent in the spectra of the cells washed with distilled water. Presumably, the ions, proteins, and amino acids present in the growth medium adsorbed on the surface of the cells would be removed in varying amounts by washing with isotonic sodium chloride solution or distilled water. It is possible that washing with water could rupture the cells and expose interior components of the cells. There were residual lines due to organic disulfide and phosphate phosphorus. These lines were assumed to be inherent to the surface structure of the cell.

The intensities of the photoelectron lines can be converted to approximate atom ratios using appropriate elemental sensitivity factors.[24]

Due to the uncertainty associated with the carbon line intensity, the oxygen 1s and nitrogen 1s line intensities are believed to be the most reliable parameters to characterize the surface of the cells. These line

Table 12. Electron Line Intensities and Atom Ratios for Chemically Fixed and Critical Point Dried Normal and Transformed[37] Cells

Sample	Core level	Intensity (1000 cps)	Corrected intensity	Oxygen/nitrogen atom ratio
7	C 1s	75		
	O 1s	53.5	85	2.6
	N 1s	14	33.3	
	Si 2p	1.5 (O 1s, 9.52)[a]		
8	C 1s	72.5		
	O 1s	58.5	76	2.0
	N 1s	15.7	37.5	
	Si 2p	3.0		
9	C 1s	61.5		
	O 1s	45	78	2.5
	N 1s	12	28.5	
	Si 2p	0.7 (O 1s, 4.5)		
10	C 1s	31		
	O 1s	16.8	27.8	2.93
	N 1s	4	9.5	
	Si 2p	0.4 (O 1s, 2.56)		
11	C 1s	55		
	O 1s	43.5	59	
	N 1s	11		2.2
	Si 2p	2.0 (O 1s, 12.7)	26.2	
12	C 1s	26		
	O 1s	16.8	32.3	2.72
	N 1s	5	11.9	
	Si 2p			

[a] Calculated on the basis of a ratio of the oxygen-to-silicon intensity ratio of 6.4. The oxygen-to-silicon ratio of 6.4 was determined from clean glass microscope coverslips.

intensities and the oxygen-to-nitrogen atom ratios are presented in Table 11.

The oxygen-to-nitrogen atom ratio varied between 2.78 and 1.95 for these three specimens. The line intensities from normal and transformed cells (unfixed preparations on polystyrene) are presented in Table 12. Wide-scan spectra from those specimens contained lines due to oxygen, nitrogen, carbon, sodium, and phosphorus. The oxygen-to-nitrogen atom ratio was 2:1 within experimental error for these two specimens. This atom ratio was not changed due to viral transformation.

9.3. Preliminary Results on Chemically Fixed Cells

Normal cells were chemically fixed with glutaraldehyde and critically dried with Freon 113. The position and assignment of the electron lines obtained from these cells (specimen 6) are given in Table 11.

Several lines were present in the spectrum of the fixed cells that were absent in the spectra of the unfixed cells. These new lines were tin (496 eV, 2p3; 488 eV 3s); lead (144 and 139 eV 4f7); silicon (154 and 103 eV, 2s and 2p); and gold (86.5 and 83.0 eV, 4f7/2, 4f5/2). The silicon lines resulted from the exposed substrate, and the gold lines originated from the sample mount surface not covered by the glass disk containing the cells. Tin and lead were found in varying amounts in the fixed samples and not in the fresh unfixed cells. These elements were apparently introduced from the reagents and solvents used in the fixing procedure.

9.4. Results on Fixed Normal and Transformed Cells

A number of normal and transformed cell preparations were surface analyzed to see if reproducible differences could be measured. These cell specimens were all grown on glass substrates and chemically fixed and critical-point-dried prior to analysis. Wide-scan spectra indicated the presence of varying amounts of tin and lead in these preparations. Figure 30 illustrates the single-scan electron lines obtained from normal cells (specimen 7, Table 1). Table 12 contains data from two different preparations of cells. For each preparation, normal and transformed cells were analyzed. The oxygen-to-nitrogen atom ratio for normal cells grown at different times was fairly reproducible. This ratio was 2.6 for specimen 7 and 2.5 for specimen 9. The corresponding ratio for the transformed cells was 2.0 for specimen 8 and 2.2 for specimen 11. These data would suggest that the oxygen-to-nitrogen ratio on the surface decreases slightly when the cells are transformed. In order to reduce the contribution of the glass substrate signals to the signal from the cells, grazing-angle measurements were made on these specimens. The relative carbon, oxygen, and nitrogen

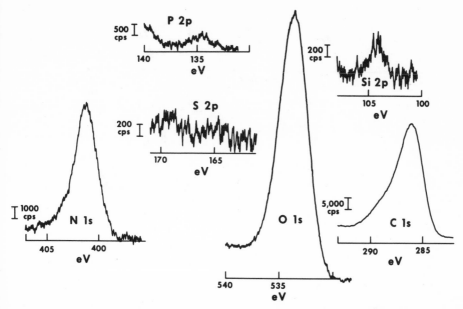

Figure 30. Single-scan electron lines from normal cells (specimen 7).[37]

signal intensities changed, and the silicon signal was virtually absent in the grazing-angle spectra. The oxygen-to-nitrogen atom ratio increased from 2.5 for sample 9 to 2.9 for sample 10 for grazing-angle measurements. This ratio for the transformed cells changed from 2.2 for sample 11 to 2.7 for sample 12 from grazing-angle measurements. The grazing-angle data indicate a higher O/N ratio in the extreme outer layers. For normal cells this ratio changes from 2.5 to 2.9. The O/N ratio decreases slightly for transformed cells. This ratio from grazing-angle data changed from 2.9 to 2.7. The grazing-angle data are in agreement with the normal-angle data in that the O/N ratio decreases slightly for the transformed cells.

9.5. Oxygen Plasma Etching and Phosphorus Composition Profile

The change in concentration of phosphorus and sodium as a function of depth into the interior of the cell was examined for both normal and transformed cells. OPE was used to etch the cells and the XPS to determine the relative concentration. For reasons previously discussed (Section 4), the rate of etching was estimated to be 9.0 nm/min and etch times were kept under 5 min. The P 2p electron line intensity as a function of etch time for normal cells is given in Figure 31.

The electron line intensity data for phosphorus, silicon, and sodium

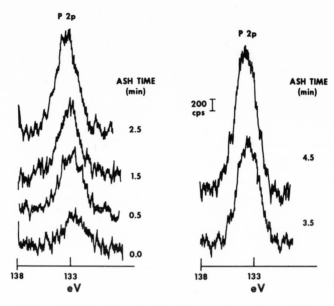

Figure 31. Phosphorus 2p lines after varying etch times.[37]

at 1 min ashing intervals up to 4.5 min for normal and transformed cells are given in Table 13.

Glass contains sodium, and the contribution of the sodium intensity from the exposed glass substrate has been estimated from the intensity of the silicon line and subtracted from the total sodium intensity. The sodium intensity data presented in Table 13 have been adjusted in this way.

A plot of the elemental composition as a function of plasma oxygen ash time yields a composition profile curve. The phosphorus composition

Table 13. Phosphorus 2p, Silicon 2p, and Na Auger (KKL 122) Line Intensities from Normal and Transformed Cells as a Function of Oxygen Plasma Oxidation[37]

Plasma exposure time (min)	Normal			Transformed		
	P 2p intensity	Si 2p intensity (10^3 cps)	Na KLL corrected intensity	P 2p intensity (10^3 cps)	Si 2p intensity (10^3 cps)	Na KLL corrected intensity
0.0	0.56	1.5	1.85	0.5	3.3	3.05
0.5	1.05	2.5	5.41	0.7	6	9.87
1.5	1.3	4.0	6.55	1.1	8	10.3
2.5	1.74	4.5	8.53	1.3	9.4	10.88
3.5	1.8	6.0	9.67	1.6	10	12.3
4.5	2.6	9.0	11.36	2.0	11.0	13.67

profile curve for normal and transformed cells at ash times up to 4.5 min is given in Figure 32. Ash time in minutes can be converted to an approximate etch depth using the conversion factor of 9.0 nm/min. As can be seen from Figure 32 the composition profile for the two types of cells is fairly similar up to ash times for 4.5 min. The phosphorus concentration increases linearly with time during first 5 min ashing. This is most reasonably interpreted as a uniform distribution of phosphorus in the region extending from the surface of the cell to well within the interior. A phosphorus-rich or -depleted region would presumably result in a change in the slope of the curve.

9.6. Estimation of Cell Coverage of Substrate

Some estimation of the coverage of the glass substrate by cells and the rate of oxidation of the cells by the oxygen plasma can be obtained from the intensity of the Si 2p electrons from the glass substrate. The initial Si 2p intensity is around 2000 counts/sec. The Si 2p electron intensity from a clean 1/4-in.-diameter glass disk is around 20,000 counts/sec. Assuming that the intensity of the Si 2p electron signal is inversely proportional to the degree of surface coverage, the culture cells cover about 90% of the glass surface. The increase in the Si 2p signal intensity upon oxygen plasma treatment is related to the rate of oxidation of the cell on the substrate by plasma oxidation. Table 13 contains the Si 2p electron intensity as a func-

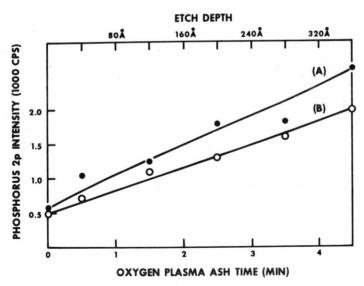

Figure 32. Phosphorus depth profile curve for normal (A) and transformed (B) cells.[37]

tion of oxygen plasma etch time for the normal and transformed cells. The Si 2p signal increases sharply during the first 1.5 min of etching and then increases slowly and steadily. The thin outer portion of the cell near the edge is apparently rapidly oxidized, exposing substrate. After oxidation of this thin exterior portion, the substrate is exposed more slowly due to coverage by a thicker portion of the cell.

9.7. Calcium Levels in Cultured Hepatoma Cells during Pyruvate-Induced DNA Synthesis

Induction of DNA synthesis is associated with increased calcium levels in cultured cells. The calcium distribution as a function of depth in the surface of hepatoma cells (HTC$_4$) was investigated using XPS and OPE.[38] The calcium composition of quiescent nonreplicating cells in minimal growth medium containing 1% serum was compared to cells in which DNA synthesis was increased 15 times by the addition of sodium pyruvate to the medium. Cells were grown on 1/4-in. glass disks washed and air dried prior to analysis. Wide-scan electron spectra are presented in Figure 33 from

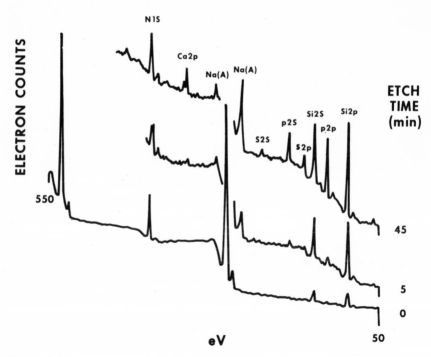

Figure 33. XPS wide-scan spectra from hepatoma cells before and after OPE for various times.[38]

unetched cells and cells etched for 5 and 45 min. These cells were grown in the presence of sodium pyruvate (1mg/ml), and their spectra are typical examples of the spectra obtained from those types of cells. The calcium signal from cells grown in sodium pyruvate was typically 2.5 times the signal obtained from control cells after 45 min OPE time. The Si 2p electron line intensity from the glass substrate was used to monitor the extent of plasma etching, and the silicon signal was often similar to within 7 parts per thousand after 45 min of plasma etching. Pyruvate-induced DNA synthesis is apparently preceded by an increase in calcium levels within the cell, but very little calcium increase was noted in the inner membrane and no increase was found in the outer membrane.

10. Red Blood Cells

10.1. Surface and Depth Profile Analysis

The red-blood-cell membrane has been extensively studied due to its availability and relative simplicity compared to other membrane systems.[108]

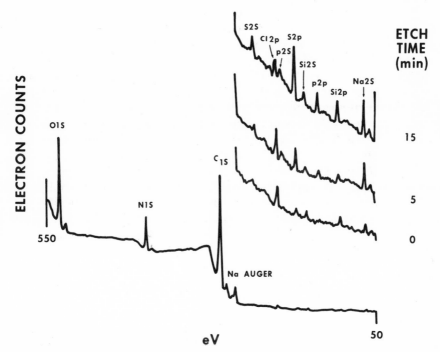

Figure 34. Wide-scan spectra of human erythrocytes as a function of OPE etch time.[109]

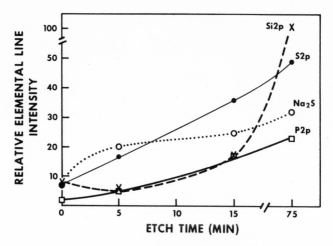

Figure 35. XPS elemental line intensities as a function of OPE time for human erythrocyte monolayers.

Meisenheimer *et al.*[35] have used argon ion etching and X-ray photoelectron spectroscopy to depth-profile elements in erythrocytes. In addition to detecting the elements C, N, O, P, S, Cl, and K on the surface, they detected thallium at a depth of about 100 Å.

Fisher and Millard[109] employed OPE and XPS to depth-profile red blood cells. Samples of closely packed monolayers of washed human erythrocytes were prepared on 1/4-in. glass disks.[110] Wide-scan spectra of human erythrocytes as a function of OPE etch time are shown in Figure 34. The intensities of photoelectron lines due to Si, S, Na, and P are given as a function of etch time in Figure 35. Previous studies on the uniformity of OPE[37,46] indicated that the etching process may become significantly nonuniform after 5–10 min etching. In this region, the Si 2p signal did not change appreciably, indicating that essentially little new substrate was exposed. The P 2p and S 2p signals increased in a fairly linear fashion. The sodium ion 2s signal increased between 0 and 5 min, then became relatively constant. Sulfur and phosphorus seem to be fairly uniformly distributed below the surface, while sodium ion may be concentrated near the surface. This may be a real phenomena or simply an artifact resulting from diffusion of sodium ion to the surface region during sample preparation.

11. References

1. D. M. Hercules, Electron spectroscopy: X-ray and electron excitation, *Anal. Chem.*, *48*, 294R–313R (1976).

2. C. A. Evans, Jr., Surface and thin film compositional analysis, *Anal. Chem., 47,* 819A (1975).
3. P. W. Palmberg, in: *Electron Spectroscopy* (D. A. Shirley, ed.), p. 835, North Holland, Amsterdam (1972).
4. R. C. Hughes, in: *Essays in Biochemistry* (P. N. Campbell and W. N. Aldridge, eds.), Vol. 2, pp. 1–36, Academic Press, New York (1975).
5. G. L. Nicolson, Transmembrane control of the receptors on normal and tumor cells. I. Cyroplasmic influence over cell surface components, *Biochim. Biophys. Acta, 457,* 57–108 (1976).
6. G. M. Edelman, Surface modulation in cell recognition and cell growth, *Science, 192,* 218 (1976).
7. J. L. Marx, Biochemistry of cancer cells. Focus on the cell surface, *Science, 183,* 1279–1282 (1974).
8. G. M. W. Cook and R. W. Stoddart, *Surface Carbohydrates of the Eukaryotic Cell,* Academic Press, New York (1973).
9. S. J. Singer and G. L. Nicolson, *Science, 175,* 720–731 (1972).
10. N. L. Warner, in: *Advances in Immunology* (F. J. Dixon and H. G. Kunkel, eds.), Vol. 19, p. 67, Academic Press, New York (1974).
11. J. J. Marchalonis, Lymphocyte surface immunoglobulins, *Science, 190,* 20 (1975).
12. John Tooze (ed.), *Molecular Biology of Tumor Viruses,* Cold Spring Harbor Laboratory, New York (1973).
13. P. Cuatrecasas, Insulin receptor of liver and fat cell membranes, *Fed. Proc., 32,* 1838 (1973).
14. A. J. Wicken and K. W. Knox, Lipoteichoic acids: A new class of bacterial antigen, *Science, 187,* 1161–1167 (1975).
15. D. M. Hercules, Electron spectroscopy, *Anal. Chem., 42,* 20A (1970).
16. W. E. Swartz, Jr., X-ray photoelectron spectroscopy, *Anal. Chem., 45,* 788A–800A (1973).
17. C. R. Brundle, Application of electron spectroscopy to surface studies, *J. Vac. Sci. Technol., 2,* 212–224 (1974).
18. D. T. Clark and W. J. Feast, Application of electron spectroscopy for chemical applications to studies of structure and bonding in polymeric systems. *Macromol. Sci. Rev. Macromol. Chem. C12,* 191–286 (1975).
19. K. Siegbahn, Electron spectroscopy, an outlook, *J. Electron Spectrosc. Relat. Phenom., 5,* 3–97 (1974).
20. T. A. Carlson, *Photoelectron and Auger Spectroscopy,* Plenum Press, New York (1975).
21. C. R. Brundle, Some recent advances in photoelectron spectroscopy, *Appl. Spectrosc., 25,* 8 (1971).
22. K. Siegbahn, D. Hammond, H. Fellner-Feldegg, and E. F. Barnett, Electron spectroscopy with monochromatized X-rays, *Science, 176,* 245 (1972).
23. C. A. Lucchesi and J. E. Lester, Electron spectroscopy instrumentation, *J. Chem. Educ., 50,* A205–A269 (1973).
24. C. A. Evans, Jr., Surface and thin film analysis, *Anal. Chem., 47,* 855A (1975).
25. C. D. Wagner, Sensitivity of detection of the elements by photoelectron spectrometry, *Anal. Chem., 44,* 1050–1053 (1972).
26. H. Berthon and C. K. Jorgensen, Relative photoelectron signal intensities obtained with a Mg X-ray source, *Anal. Chem., 47,* 482 (1975).
27. M. Vulli, M. Janghorbani, and K. Starke, Intra elemental photoelectron line intensities and their significance to quantitative analysis, *Anal. Chim. Acta, 82,* 121–135 (1976).
28. D. T. Clark, W. J. Feast, W. K. R. Musgrave, and I. Ritchie, Applications of ESCA to polymer chemistry. Part VI. Surface fluorination of polyethylene application of

ESCA to examination of structure as a function of depth, *J. Polym. Sci. Polym. Chem. Ed.*, *13*, 857–890 (1975).

29. J. D. Andrade, University of Utah, personal communication (1976).
30. C. S. Fadley, Solid state and surface analysis by means of angular dependent XPS, *Prog. Solid State Chem.*, *2*, 265–343 (1976).
31. C. Lederer, Lawrence Berkeley Laboratory, Berkeley, California, personal communication (1973).
32. C. S. Fadley, Core and valence electronic states studied with X-ray photoelectron spectroscopy, PhD thesis, Lawrence Radiation Laboratory Report UCRL 19535 (1970).
33. T. Hall, P. Echlin, and R. Kaufmann, *Microprobe Analysis as Applied to Cells and Tissues*, Academic Press, New York (1974).
34. J. Baddiley, I. C. Hancook, and P. M. A. Sherwood, X-ray photoelectron studies of Magnesium ions bound to the cell walls of gram-positive bacteria, *Nature*, *243*, 43 (1973).
35. R. G. Meisenheimer, J. W. Fischer, and S. J. Rehfeld, Thallium in human erythrocyte membranes: An X-ray photoelectron spectroscopic study, *Biochem. Biophys. Res. Commun.*, *68*, 994 (1976).
36. D. T. Clark, J. Peeling, and L. Colling, An experimental and theoretical investigation of the core level spectra of a series of amino acids dipeptides and polypeptides, *Biochim. Biophys. Acta*, *453*, 533–545 (1976).
37. M. M. Millard and J. C. Bartholomew, Application of XPS to surface studies of mammalian cells grown in culture, *Anal. Chem.*, *49*, 1290–1296 (1977).
38. L. Pickart, M. M. Millard, B. Beiderman, and M. M. Thaler, Surface analysis and depth profiles of Ca in Hepatoma cells during pyruvate induced DNA synthesis, *Fed. Proc.*, *36*, 703 (1977).
39. D. L. White, S. B. Andrews, J. W. Faller, and R. J. Barnett, Chemical nature of osmium tetraoxide fixation and staining of membranes by XPS, *Biochim. Biophys. Acta*, *436*, 577–592 (1976).
40. G. K. Wehner, in: *Methods and Phenomena. Vol. 1. Methods of Surface Analysis* (A. W. Czanderna, ed.), pp. 5–37, Elsevier, New York (1975).
41. R. J. Colton and W. J. Rabalais, Electronic structure of tungsten and some of its borides, carbides, nitrides and oxides by X-ray photoelectron spectroscopy, *Inorg. Chem.*, *15*, 236–238 (1976).
42. K. S. Kim, W. E. Baitinger, J. W. Amy, and N. J. Winograd, ESCA studies of metal oxygen surfaces using Argon and oxygen ion bombardment, *J. Elec. Spectrosc. Relat. Phenom.*, *5*, 351 (1974).
43. R. S. Thomas, in: *Techniques and Applications of Plasma Chemistry* (J. R. Hollahan and A. T. Bell, eds.), p. 255, John Wiley & Sons, New York (1974).
44. R. S. Thomas and J. R. Hollahan, *Scanning Electron Microscopy Proceedings of the 7th Annual Scanning Electron Microscopy Symposium, Part I*, p. 83, ITT Research Institute, Chicago, Illinois (1974).
45. J. R. Hollahan, in: *Techniques and Applications of Plasma Chemistry* (J. R. Hollahan and A. T. Bell, eds.), p. 246, John Wiley & Sons, New York (1974).
46. R. S. Thomas, M. M. Millard, and R. Scherrer, in: *Electron Microscopy Society of America 34th Annual Proceedings* (G. W. Bailey, ed.), pp. 134–135, Miami Beach, Florida (1976).
47. M. M. Millard, in: *Protein–Metal Interactions* (M. Friedman, ed.), pp. 589–619, Plenum Press, New York (1974).
48. M. V. Zeller and R. G. Hayes, X-ray photoelectron spectroscopic studies on electronic structures of porphyrin and phthalocyanine compounds, *J. Am. Chem. Soc.*, *95*, 3855–3860 (1973).

49. Y. Niwa, H. Kobayashi, and T. Tsuchiya, X-ray photoelectron spectroscopy of tetraphenylprophine and phthalocyanine, *J. Chem. Phys.*, 60, 799–807 (1974).

50. Y. Niwa, H. Kobayashi, and T. Tsuchiya, X-ray photoelectron spectroscopy of A2a porphyrins, *Inorg. Chem.*, 13, 2891–2895 (1974).

51. S. C. Khandelwal and J. L. Roebber, Photoelectron spectra of tetraphenylporphine and some metallotetraphenylporhyrins, *Chem. Phys. Lett.*, 34, 355–359 (1974).

52. Y. Niwa, X-ray photoelectron spectra of reduced porphines, *J. Chem. Phys.*, 62, 737–738 (1975).

53. D. H. Karweik and N. Winograd, Nitrogen charge distributions in free base porphyrins, metalloprophyrins and their reduced analogues observed by X-ray photoelectron spectroscopy, *Inorg. Chem.*, 15, 2336–2342 (1976).

54. J. A. Signorelli and R. G. Hayes, Studies of tetra-(*p*-fluorophenyl) porphyrins by X-ray photoelectron spectroscopy. Effect of reduction on charge distribution in Fe compounds, *J. Chem. Phys.*, 64, 4517 (1976).

55. U. Wesser, W. Paschen, and M. Younes, Singlet oxygen and superoxide dismutase cuprein, *Biochem. Biophys. Res. Commun.*, 66, 769–777 (1975).

56. U. Weser, G. J. Strobel, and W. Voelter, Carbon-13 NMR and X-ray photoelectron spectroscopy of copper AMP, *Fed. Eur. Biochem. Soc. Lett.*, 41, 243–247 (1974).

57. P. T. Andrews, C. E. Johnson, B. Wallbank, R. Cammack, D. O. Hall, and K. K. Rao, X-ray photoelectron spectra of iron sulfur proteins. *Biochem. J.* 149, 471–474 (1975).

58. G. Sokolowski, W. Pilz, and U. Weser, X-ray photoelectron spectral properties of Hg thionein, *Fed. Eur. Biol. Soc. Lett.*, 48, 222–225 (1974).

59. A. Wendel, W. Pilz, R. Ladenstein, G. Sawatzki, and U. Weser, Substrate induced redox change of selenium and glutathione peroxidase studies by X-ray photoelectron spectroscopy, *Biochim. Biophys. Acta*, 377, 211–215 (1975).

60. D. Chiu, A. L. Tappel, and M. M. Millard, X-ray photoelectron spectroscopy of selenium glutathione peroxidase from rat liver, *Arch. Biochem. Biophys.* 184, 209–214 (1977).

61. G. Sokolowski and U. Weser, Formation and circular dichroism and X-ray photoelectron spectra of hepatic zinc thionein, *Hoppe-Seyler's Z. Physiol. Chem.*, 356, 1715–1726 (1975).

62. N. Winogard, A. Shepard, D. H. Karweik, V. J. Doester, and F. K. Fong, X-ray photoelectron spectroscopy of thermal stability of chlorophyll A monohydrate spinach, *J. Am. Chem. Soc.*, 98, 2369–2370 (1976).

63. C. R. Cothern, W. E. Moddeman, R. G. Albridge, W. J. Sanders, P. L. Kelley, W. S. Hanley, and L. Field, Bonding properties of D-penicillamine and related compounds by X-ray photoelectron spectroscopy, *Anal. Chem.*, 48, 162–166 (1976).

64. H. Falk, O. Hofer, and H. Lehner, Chemistiry of pyrrole pigments. 2. X-ray photoelectron spectroscopy of some pyrrole pigments, *Monatsh. Chem.*, 105, 366–378 (1974).

65. H. Rupp and U. Weser, X-ray photoelectron spectroscopy of some selenium containing amino acids, *Bioinorg. Chem.*, 5, 21–32 (1975).

66. D. M. Derrier and R. R. Benerito, Mono- and diquaternary ammonium cellulose cottons prepared in nonaqueous media, *J. Appl. Polym. Sci.*, 19, 3211–3220 (1975).

67. D. M. Soignet, R. J. Berni, and R. R. Benerito, Electron spectroscopy for chemical analysis (ESCA). A tool for studying treated textiles. *J. Appl. Polym. Sci.*, 20, 2483–2495 (1976).

68. D. M. Soignet, R. J. Berni, and R. R. Benerito, ESCA of ThPOH-NH$_3^+$ treated Fabrics, *Text. Res. J.*, 45, 28 (1975).

69. D. M. Soignet, W. J. Connick, and R. R. Benerito, Use of ESCA in a study of oil-repellent finishes, *Text. Res. J.*, 46, 611–616 (1976).

70. M. M. Millard, Analysis of surface oxidized wool fiber by X-ray electron spectros-copy, *Anal. Chem.*, *44*, 828–829 (1972).

71. M. M. Millard, K. S. Lee, and A. E. Pavlath, Continuous low-temperature discharge treatment of wool yarn. A process for making wool yarn oil repellent and shrink resistant, *Text. Res. J.*, *42*, 307–313 (1972).

72. M. M. Millard and A. E. Pavlath, Surface analysis of wool fibers and fiber coatings by X-ray photoelectron spectroscopy, *Text. Res. J.*, *42*, 460–463 (1972).

73. M. M. Millard, in: *Characterization of Polymer Surfaces* (L. H. Lee, ed.), Vol 2, p. 86, Academic Press, New York (1977).

74. M. M. Millard, J. J. Windle, and A. E. Pavlath, Plasma synthesis of fluorocarbon films, *J. Appl. Polym. Sci.*, *17*, 2501–2507 (1973).

75. A. E. Pavlath, K. S. Lee, and W. M. Ward, Film formation from insoluble proteins, Symposium on membranes, 1 North American Chemical Congress, Mexico City, Mexico (1975) MACRO 20 Abstract.

76. U. Gelius, P. F. Heden, J. Hedman, B. J. Lindberg, R. Maune, R. Nordberg, C. Nordling, and K. Siegbahn, Molecular spectroscopy by means of ESCA, Part 2, Carbon compounds, *Phys. Scr.*, *2*, 70–80 (1970).

77. J. B. Donnet, H. Dauksch, J. Escard, and C. Winter, Surface oxidation of carbon fibers studied by X-ray photoelectron spectroscopy, *C.R. Acad. Sci.*, *275C*, 1219 (1972).

78. B. J. Lindberg, K. Hamrin, G. Johansson, U. Gelius, A. Fahlman, C. Nordling, and K. Siegbahn, Molecular spectroscopy by means of ESCA, Part 1, Sulfur compounds, *Phys. Sci.*, *1*, 286–298 (1970).

79. J. L. Ogilvie and A. Wolbery, An internal standard for electron spectroscopy for chemical analysis studies of supported catalysts, *Appl. Spectrosc.*, *26*, 401 (1972).

80. G. C. Allen and P. M. Tucker, Surface oxidation of uranium metals studied by XPS, *J. Chem. Soc. Dalton Trans.*, *5*, 470 (1973).

81. K. S. Kim and N. Winograd, Observation of polymorphic lead monoxides surfaces using XPS, *Chem. Phys. Lett.*, *19*, 209 (1973).

82. F. J. Grunthaner, Electronic structure surface reactivity and site analysis of transition metal complexes and metalloproteins by X-ray photoelectron spectroscopy, PhD thesis, California Institute of Technology, Pasadena, California (1974).

83. M. Barber, E. L. Evans, and J. M. Thomas, Oxygen chemisorption on basal faces of graphite, an XPS study, *Chem. Phys. Lett.*, *18*, 423 (1973).

84. M. P. Klein and L. N. Kramer, in: *Symposium: Seed Proteins*, pp. 265–276, Avi Publishing Co., Westport, Connecticut (1976).

85. J. Peeling, D. T. Clark, I. M. Evans, and D. Boulter, Evaluation of the ESCA technique as a screening method for estimation of protein content and quality in seed meals, *J. Sci. Food Agr.*, *27*, 331–340 (1976).

86. R. Nordberg, R. G. Albridge, T. Bergmark, U. Erickson, J. Hedman, C. Nordling, K. Seigbahn, and B. J. Lindberg, Molecular spectroscopy by means of ESCA, Charge distribution in nitrogen compounds, *Ark. Kemi*, *28*, 257–278 (1968).

87. M. M. Millard, in: *Techniques and Application of Plasma Chemistry* (A. T. Bell and J. R. Hollahan, eds.), pp. 177–214, John Wiley, New York (1974).

88. A. G. Pittman, in: *High Polymers* (L. A. Wall, ed.), Vol. 25, pp. 419–449, John Wiley, New York (1972).

89. A. G. Pittman, W. L. Wasley, and J. N. Roitman, Fluorosiloxane copolymers as soil release agents, *Text. Chem. Color.*, *4*, 278–333 (1972).

90. D. T. Clark, W. J. Feast, D. Kilcast, and W. K. R. Musgrave, Applications of ESCA to polymer chemistry, III, structures and bonding in homopolymers of ethylene and fluoroethylenes and determination of the compositions of fluorocopolymers, *J. Polym. Sci., Polym. Chem. Ed.*, *11*, 389 (1973).

91. R. W. Phillips, Quantitative X-ray photoelectron spectroscopic analysis of fluoro-chemicals, *J. Colloid Interface Sci., 47,* 687 (1974).

92. M. M. Millard and A. E. Pavlath, Surface analysis of plasma polymerized fluorocarbon films by x-ray photoelectron spectrometry, *J. Macromol. Sci. Chem., A10* (3), 579–597 (1976).

93. D. T. Clark, D. Kilcast, W. J. Feast, and W. K. R. Musgrave, Application of ESCA to polymer chemistry: Studies of some nitroso rubbers, *J. Polym. Sci., Polym. Chem. Ed., 10,* 1637 (1972).

94. M. M. Millard, Formation of fluorocarbon films by ultraviolet surface photopolymerization, *J. Appl. Polym. Sci., 18,* 3319–3325 (1974).

95. M. M. Millard and M. S. Masri, Detection and analysis of protein functional group modification by X-ray photoelectron spectrometry, *Anal. Chem., 46,* 1820 (1974).

96. M. M. Millard and M. Friedman, XPS of BSA and ethyl vinyl sulfone modified BSA, *Biochem. Biophys. Res. Commun., 70,* 445–451 (1976).

97. J. Mandelstam and K. McQuillen (eds.), *Biochemistry of Bacterial Growth,* John Wiley, New York (1975).

98. D. C. Birdsell, R. J. Doyle, and M. Morgenstern, Organization of teichoic acid in the cell wall of Bacillus subtilis, *J. Bacteriol., 121,* 726–734 (1975).

99. B. D. Davis, R. Dublecco, H. N. Eisen, H. S. Ginsberg, W. B. Wood, Jr., and M. McCarty, *Microbiology,* 2nd ed., p. 116, Harper and Row, New York, (1973).

100. O. Luderitz, Recent results on biochemistry of cell wall L polysaccharide of *Salmonella* bacteria, *Angew. Chem., 9,* 649 (1970).

101. M. M. Millard, R. Scherrer, and R. S. Thomas, Surface analysis and depth profile composition of bacterial cells by X-ray photoelectron spectroscopy and oxygen plasma etching, *Biochem. Biophys. Res. Commun., 72,* 1209 (1976).

102. M. K. Nemanic, On cell surface labeling for the scanning electron microscope, in: *Scanning Electron Microscopy, Part 1, Proceedings of the Eighth Annual SEM Symposium* (O. Johari and I. Corvin, eds.), p. 341, ITT Research Institute, Chicago, Illinois (1975).

103. T. Hall, P. Echlin, and R. Kaufmann (ed.), *Microprobe Analysis as Applied to Cells and Tissues,* Academic Press, New York (1974).

104. A. Boyde, in: *Scanning Electron Microscopy, Part II, Proceedings of Workshop on Biological Specimen Preparation for SEM* (O. Johari and I. Corvin, eds.), pp. 257–264, ITT Research Institute, Chicago, Illinois (1972).

105. A. Boyde and P. Vesely, in: *Scanning Electron Microscopy, Part II, Proceedings of Workshop on Biological Specimen Preparation for SEM* (O. Johari and I. Corvin, eds.), pp. 265–271, ITT Research Institute, Chicago, Illinois (1972).

106. A. Boyde, R. A. Weiss, and P. Vesely, Scanning electron microscopy of cells in culture, *Exp. Cell Res., 71,* 313–324 (1972).

107. O. Johari and I. Corvin (eds.), *Scanning Electron Microscopy Symposium Proceedings,* ITT Research Institute, Chicago, Illinois (1975) and preceding volumes.

108. V. T. Machesi, H. Furthmayr, and M. Tomita, in: *Annual Review of Biochemistry* (E. E. Snell, ed.) Vol. 45, pp. 667–698, Annual Reviews Inc., Palo Alto, California (1976).

109. K. A. Fisher and M. M. Millard (1976), unpublished data.

110. K. A. Fisher, Analysis of membrane halves, cholesterol, *Proc. Natl. Acad. Sci. U.S.A., 73,* 173–177 (1976).

111. K. W. Knox and A. J. Wicken, Immunological properties of teichoic acids, *Bacteriol. Rev. 37,* 227 (1973).

112. A. L. Lehninger, *Biochemistry,* Second ed., p. 269, Worth Publishers, New York (1976).

2

Surface Analysis using Energetic Ions

J. A. Borders

1. Introduction

The surface and near-surface properties of solid materials are of great importance in modern technology. Whereas the surface and near-surface properties of a solid may not be as predictable from a theoretical standpoint, nor as easy to calculate from a numerical standpoint, as the bulk properties of a solid, they affect the manner in which that material interacts with the outside world. Material properties, such as catalytic behavior, reactivity, and resistance to corrosion, and frictional wear are all determined in large part by the properties of surfaces. Realization of the importance of surface properties has led to the development of a great variety of surface characterization techniques. Many of these modern techniques are relatively well known and widely used; Auger spectroscopy and electron microprobe analysis are two such techniques. A class of techniques whose capabilities are not as widely appreciated utilizes energetic ions to analyze surface layers of materials. These energetic ion analysis techniques all depend on the interactions between a beam of energetic ions and the atoms of the target material being studied, which until quite recently were the exclusive domain of the low-energy nuclear and atomic physicists.

Because ion–solid interactions are quite different from electron–solid and photon–solid interactions, the data obtained using energetic ion analysis are quite different from that obtained using electron- or photon-excited surface analysis methods. The data are, however, quite complementary to those obtained using other techniques and yield information which may be difficult or impossible to obtain using other methods.

J. A. Borders • Sandia Laboratories, Albuquerque, New Mexico 87185.

In this chapter an introduction into the general features of ion analysis will be presented, and each of the three most widely used techniques will be reviewed in some detail. Examples will be presented which illustrate the unique features of the various techniques and, finally, the techniques will be compared with other modern methods of surface analysis.

2. General Features of Energetic Ion Analysis

Modern analysis techniques utilize a wide variety of different physical phenomena to obtain information about a solid surface. These phenomena are generated by incident particles which may be photons, electrons, ions, or any convenient particle which interacts with the target material of interest. It is by examining the scattered or emitted particles that each technique is able to obtain information about the target. In order to properly interpret the data which are obtained in such an experiment, a knowledge of how the incident particles interact with the target is necessary.

All of the energetic ion techniques to be discussed in this chapter depend on an interaction between an incident ion and the atoms of the target. The incident ion energies are generally in the range 100 keV to 2.0 MeV. Until recently, the study and use of these interactions were strictly in the realm of low-energy nuclear and atomic collision physicists who were concerned with the interactions themselves rather than the information which could be obtained about a solid target. The individual interactions will be discussed in more detail in later sections of this article, but, briefly, ion backscattering relies on elastic scattering of the incident ions at large angles to provide information on the mass of the target atom from which the ion was scattered. Similarly, in ion-induced X-ray analysis, characteristic X rays are detected from the target atoms which were excited by the incident ions; in nuclear microanalysis, the products of inelastic nuclear interactions between the incident ions and a specific isotope of a target species are detected.

The physical concepts behind these various energetic ion analysis techniques are not new. However, the use of these techniques in comprehensive research programs has been almost entirely confined to the last decade, with a majority of the work having been initiated after 1970. Three international conferences devoted to ion-beam surface-layer analysis have been held with a fourth planned for the summer of 1979. The proceedings[1-3] of the first three conferences in this series are in themselves the best sources of background information. Additionally, there are a number of excellent review articles[4-6] which cover some or all aspects of the field.

One of the useful features of energetic ion analysis, particularly ion backscattering, is the capability of obtaining information about the depth distribution of atomic composition. This is possible because: (1) an ion loses a large amount of energy to the electronic system while passing the atoms of the target as compared to the energy resolution with which an ion can be detected; (2) the trajectories of the ions which penetrate the solid without being scattered by target nuclei are straight; and (3) the penetration and escape depths for energetic ions in solids are relatively large (1–10 μm).

A unique feature of energetic ion analysis is that the depth-profiling capabilities mentioned above are relatively nondestructive. Most surface analysis techniques obtain depth information by making one measurement at the sample surface and then removing a thin layer of the sample surface and making another measurement. Low-energy ion sputtering is commonly used for layer removal, and this technique destroys the sample in the process of obtaining the depth-profile information. There are also problems with sputtering causing changes in the near-surface composition of materials. The energetic ion techniques obtain the information in one measurement and do not destroy the sample. There is, of course, some damage to the material, and care must be taken to ensure that this damage does not affect the measurement.

The equipment required for these analytical techniques is rather complex and expensive. However, with small modifications, the same equipment can generally be used for all three techniques, viz., backscattering, ion-induced X rays, and nuclear reaction analysis. A schematic diagram of a typical experimental arrangement can be seen in Figure 1. An ion accelerator (often a Van de Graaff) is used to generate a beam of ions of a well-defined energy. This beam is deflected in a magnetic analyzer which selects particles of a given momentum. The beam is directed on the sample for analysis after the angular spread and lateral dimensions of the beam have been defined by an appropriate set of collimators. Collimation is particularly important in the case of channeling experiments. Channeling is a phenomenon which occurs when the incident ion beam is very nearly aligned with a particular crystal axis or plane in a single-crystal target. When these conditions are satisfied, most of the incident ions are gently guided through the crystal and the ions are prevented from approaching the target atoms very closely. This lowers the yield of close-encounter processes such as energetic ion–atom interactions. Channeling techniques can be combined with any of the energetic ion analysis methods which are being discussed in this chapter to yield additional information about single-crystal materials such as the amount of lattice disorder and the crystallographic location of impurities. However, we shall not attempt a discussion of channeling within the limited scope of this article, and the reader is referred to other sources.[7,8]

The target itself is connected to a current measuring device so that the beam current may be monitored and integrated. In order that this

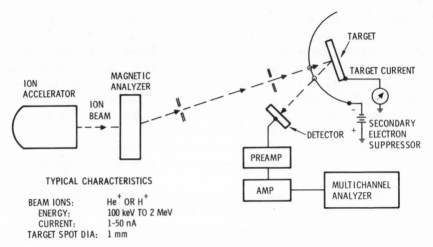

Figure 1. Experimental arrangement for energetic ion surface analysis.

current measurement be correct, some device for suppressing secondary electrons is customary. In Figure 1, this is shown as a negatively biased shield, although a Faraday cup arrangement is often used instead.

The products of a scattering event between an incident ion and a target atom are detected in a device selected for the type of particles expected. For ion backscattering and nuclear reaction microanalysis, this is generally a partially depleted, silicon surface-barrier detector. Generally, the cross sections for nuclear reactions are much smaller than for backscattering, and the emitted particle energies are much greater than the incident ion energies due to a positive Q value for many of the nuclear reactions commonly used for microanalysis. Thus for nuclear reactions, special detectors with larger depletion depths may be necessary, and thin films may be used in front of these detectors to stop ions which have been backscattered. For ion-induced X rays, the detector is usually a Ge(Li) or Si(Li) solid-state detector, although there is some use of specially designed ionization chamber detectors.[9]

The signals from the detector are amplified and shaped using standard nuclear pulse electronics. The processed pulses, whose heights are proportional to the energy of the incident particle or X ray, are then input to a multichannel analyzer. This instrument sorts the pulses into a spectrum which consists of the number of particles/X rays (on the vertical axis) detected at a particular energy vs. channel number (horizontal axis) which is directly proportional to particle or X-ray energy. These plots are known as pulse-height spectra.

The lateral extent of the beam is usually about 1 mm². Within this area, there is no lateral resolution. This lack of lateral resolution is a major disadvantage of the ion analysis techniques and is indeed a good argument for combining ion backscattering with a technique which yields lateral compositional information. This is not an inherent limitation as ion beams can be focused to achieve smaller lateral dimensions. Some work has begun at Harwell using proton "microbeams" a few micrometers across.[10]

The magnitude of the ion-beam current and the total ion dose required for an analysis will vary widely from experiment to experiment. The experimental set-up and the target to be analyzed will dictate these parameters in many situations. Typical beam currents for backscattering and X-ray analysis experiments are 1–50×10^{-9} A, and total doses for an analysis can run from 0.01 μC to 100 μC. One microcoulomb is equivalent to 6.25×10^{14} singly charged ions or the equivalent of about half a monolayer.

Finally, before proceeding to a detailed discussion of each technique, a word needs to be said about the definition of a surface. Many purists contend, and properly so, that surface refers only to the outermost layer of atoms of a material. Thus, a surface analysis technique should refer to a technique which examines only this outer layer. Most analysis techniques are sensitive to atoms below this outer layer, and, indeed, the techniques described here are sensitive to atoms from the surface to as much as 10 μm deep. When we use the term surface analysis, we will be referring to the analysis of layers near the surface but comprising more than one atom layer.

3. Energetic Ion Backscattering

Perhaps the most generally useful of all the energetic ion-beam analysis techniques is ion backscattering. Also known as Rutherford backscattering analysis or elastic backscattering analysis, this technique provides quick and easily interpreted measurements of the depth distribution of atomic composition in the near-surface layer of a solid. The advantages of the technique are many, but three characteristics of this analysis tool are particularly useful. First, the compositional information is absolute and does not depend on "standard" samples. The second unique feature of energetic ion backscattering is that it can provide a nondestructive measurement of the depth distribution of atomic composition. Finally, once the equipment is set up, the measurements are very easy to make, and data analysis is particularly simple. In this section, we will briefly discuss the physics of ion backscattering before proceeding to look at some specific examples of the use of the technique.

Elastic collisions between the incident and target atom nuclei are the interactions which provide information about the composition of the surface layer in an ion backscattering experiment. The essential features of an elastic collision are depicted schematically in Figure 2. Since the collisions are elastic, energy is conserved, and if we apply conservation of momentum as well as energy, we find that the energy of the backscattered ion is simply proportional to the incident energy with a proportionality constant which depends only on the projectile mass, the target mass, and the scattering angle θ_{lab}. The situation depicted in Figure 2 is the common situation where the target atom mass is greater than the projectile mass. One can perform analyses with the projectile mass equal to or greater than the target atom mass, but the experimental situation becomes more difficult as there are no scattered particles with $\theta_{lab} > 90°$. The important point is that the energy of the scattered ion is dependent only on the projectile and target masses and the scattering angle. In the discussion which follows, it will be implicitly assumed that the incident ion or projectile mass is less than the mass of the lightest target atom.

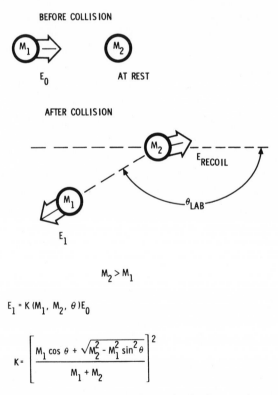

Figure 2. Schematic illustration of an elastic (Rutherford) scattering event with appropriate equations.

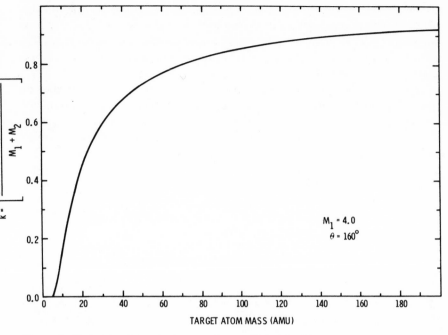

Figure 3. Variation of the elastic scattering coefficient k with target atom mass for scattering of ^{4}He particles at a laboratory scattering angle of 160°.

Ion backscattering experiments are normally performed with a specific projectile ion, and the detector normally is at a fixed scattering angle so that the constant of proportionality is a function only of the mass of the target atom from which the projectile ion was scattered. Ions which have backscattered from heavy atoms retain more of their incident energy than ions which have backscattered from lighter target atoms. To give the reader a feeling of this relationship, the constant of proportionality, k, is plotted in Figure 3 as a function of the mass of the target atom assuming an incident ^{4}He beam and a detector at a laboratory angle of 160°. In general, k is defined as

$$k = \left[\frac{M_1 \cos \theta + (M_2^2 - M_1^2 \sin^2 \theta)^{1/2}}{M_1 + M_2} \right]^2 \tag{1}$$

where M_1 and M_2 are the masses of the projectile and target atoms, respectively, and θ is the laboratory scattering angle. As can be seen from Figure 3, the mass resolution (dk/dM) is much better for light target atoms than for heavy target atoms. It should also be remembered that the scattered energy in any given situation is proportional to the incident energy. Thus, more mass resolution can be obtained for a given energy resolution merely by increasing the incident beam energy. It should be pointed out at this juncture that ion backscattering measures the mass of target atoms, not what elements they are. In many cases, this distinction is irrelevant, but it remains a fact which must be kept in mind when interpreting backscattering spectra.

If ion backscattering is to be a quantitative technique, the number or yield of scattered ions from a given target atom for a given number of incident ions must be known. This information is contained in the cross section for elastic scattering for the given ion–target combination. For most cases of interest in energetic backscattering, the scattering cross section is known to obey the Rutherford formula which is usually written[11] in center-of-mass system variables as

$$\left.\frac{d\sigma}{d\Omega}\right|_{cm} = 1.296 \left(\frac{Z_1 Z_2}{E}\right)^2 \left(\frac{M_1 + M_2}{M_2}\right)^2 \sin^{-4}\left(\frac{\theta_{cm}}{2}\right) \quad [mb/sr] \qquad (2)$$

Z_1 and Z_2 are the atomic numbers of the incident ion and the target atom, respectively, and M_1 and M_2 are their respective masses. E is the energy of the incident ion, and θ_{cm} is the center-of-mass scattering angle. When the cross section is expressed in terms of the laboratory scattering angle, θ_{lab}, it is usually expanded in the form

$$\left.\frac{d\sigma}{d\Omega}\right|_{lab} = 1.296 \left(\frac{Z_1 Z_2}{E}\right)^2 \left[\sin^{-4}\left(\frac{\theta_{lab}}{2}\right) - 2\left(\frac{M_1}{M_2}\right)^2 + \ldots\right] \quad [mb/sr] \qquad (3)$$

although it can also be written in closed form: $(d\sigma/d\Omega)$ is expressed in units of millibarns per steradian where 1 barn $= 10^{-24}$ cm^2. Several points are evident from the above equations. The most important point is that the cross section is proportional to the atomic number of the target nucleus squared. This provides ion backscattering with a very high sensitivity for heavy impurities, particularly if they are on the surface of a light substrate. The cross section is inversely proportional to the square of the incident ion energy which provides increased sensitivity at lower energies. This is normally not a great consideration, however, since the energy dependence of the stopping power must also be considered as well as the proper energy to separate the masses of the various target atoms. Another important aspect is that the cross section is only a function of the number of the protons on the nucleus and the nuclear mass; it does not depend on the atomic environment of the nucleus. Thus backscattering measurements are free of the "matrix effects" which make absolute measurements so difficult using other techniques such as electron spectroscopy.

As an illustration of the energy at which ions are scattered from various mass target atoms and the varying sensitivity to atoms of different atomic number, a hypothetical spectrum is shown in Figure 4 of a very

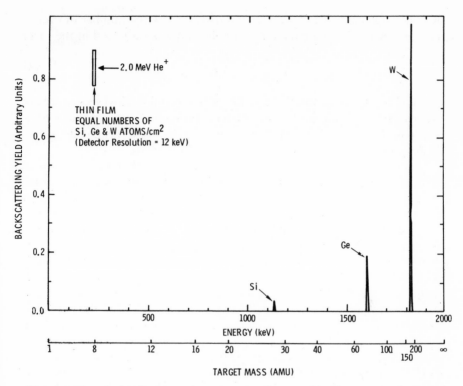

Figure 4. Hypothetical backscattering spectrum of a very thin film composed of equal numbers of Si, Ge, and W atoms.

thin film composed of equal numbers of Si, Ge, and W atoms as analyzed using 2.0 MeV He$^+$ at a laboratory scattering angle of 160°. It is assumed that the film is self-supporting and that particles which do not scatter from film atoms do not scatter from anything behind the film. As can be seen, the spectrum consists of three narrow Gaussian peaks. The energy of each of these peaks corresponds to the energy of ions scattered from the film atoms of different masses. These energies can be calculated by use of equation (2). The width of these Gaussian peaks is determined by the energy resolution of the experimental system. It has been assumed, for simplicity, that each element consists of only one isotope. The effect of the scattering cross section is quite evident in Figure 4 in that the scattering yield from the W atoms is $(74/14)^2 \sim 28$ times the scattering yield from an equal number of Si atoms.

The above discussion assumed that the incident ions did not lose any energy in passing through the target film. Actually ions lose energy to electronic processes as they pass target atoms. Each event involves the loss of a very small amount of energy, and there are a large number of events. The energy loss can be treated statistically, and it is expressed in the form

$$\frac{dE}{dx} = -N\,\epsilon(E) \qquad (4)$$

where N is the atomic density and $\epsilon(E)$ is the stopping cross section. Many experimental measurements of $\epsilon(E)$ have been made for the most commonly used ions, H^+ and $^4He^+$. For $^4He^+$ the data have been collected by Ziegler and Chu[12] into a very convenient tabular form.

In samples of finite thickness, the energy-loss mechanism to electronic processes manifests itself in the backscattering spectra themselves. In order to see how the electronic-loss mechanism affects the measured spectra and enables the experimentalist to determine the depth distribution of atomic composition, we will first consider a thin film of only one atomic species, as illustrated schematically in Figure 5. As the beam of energy E_0 strikes the film surface, some ions are elastically scattered at a given laboratory angle. These ions have energy kE_0 where k is given by equation (1). As an example, for 2 MeV $^4He^+$ ions scattering at 160° from a tungsten film, the mass of tungsten is ~184 and $k = 0.919$. The energy of the helium ions backscattered at 160° from the tungsten surface is 0.919×2000 keV = 1838 keV. The total elastic scattering cross section, however, is very small, so the unscattered beam is essentially unattenuated. Ions are backscattered from target atoms at all depths in the film, and those ions which penetrate through the film are assumed to not interact further. As shown in Figure 5, at a depth of x_1, ions are backscattered at an energy of $kE_0 - Nx_1[S_1]$, where $[S_1] = k\,\epsilon(\bar{E}_{in}) + \sec\theta\cdot\epsilon(\bar{E}_{out})$, and $\epsilon(\bar{E}_{in})$ and $\epsilon(\bar{E}_{out})$ are the stopping cross sections at average energies for the ions traversing x_1 before and after the elastic scattering event at depth x_1, respectively. The ions scattered from a small increment in depth δx_1, thus give rise to a contribution in the scattering yield at an energy $kE_0 - Nx_1[S_1]$ and in an energy interval δE_1. This is shown as a shaded band in the scattering yield at the bottom of Figure 5. A similar correlation is shown in Figure 5 for the scattering in an energy interval δE_2 about an energy $kE_0 - Nx_2[S_2]$ for scattering from target atoms in a depth interval δx_2 at a depth x_2. What this illustrates is that for a monoatomic target there is a one-to-one correspondence between the depth of the target atoms and the energy of the ions scattered from them. If ΔE is the difference between the energy of the particles scattered from the surface and at some depth x, then

$$\Delta E = Nx[S] \qquad (5)$$

which is the basic equation for determining depth using ion backscattering.

If we consider the effect of electronic energy loss on the spectrum shown in Figure 4, as we let the thin film become thicker, the narrow

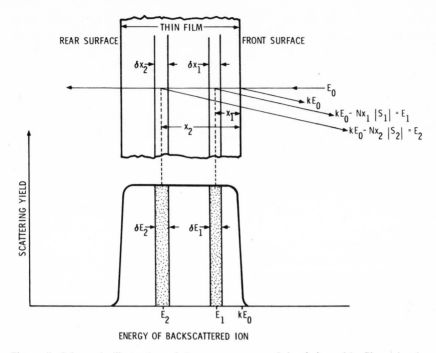

Figure 5. Schematic illustration of the measurement of depth in a thin film using ion backscattering.

Gaussian scattering peaks shown in Figure 4 become wider. Essentially, the spectrum becomes the sum of many Gaussians from many very thin layers. If the peaks become wide enough in energy, the scattering yields from different mass atoms will overlap.

Figure 6 illustrates in a schematic fashion the spectra which would be expected if the very thin film of Figure 4 were allowed to become a film ~1000 Å thick (shaded peaks) or to become so thick that the analyzing ions could not penetrate, scatter, and return through the layer to the detector (>1 μm). It is easily seen from Figure 6 that if the scattering yields from the various components of a multiatomic thin film are separated in energy, they are easily measured but that once the scattering yields from different species begin to overlap, the lighter mass components become more difficult to measure.

Figure 6 also shows a gradual increase in yield at low energies. This change is due to: (1) variation of the Rutherford cross section with energy, and (2) variation of the electronic stopping power with energy. For He ions in the energy range 1–2 MeV, the first effect is predominant because the electronic stopping power does not vary a great deal as a function of energy

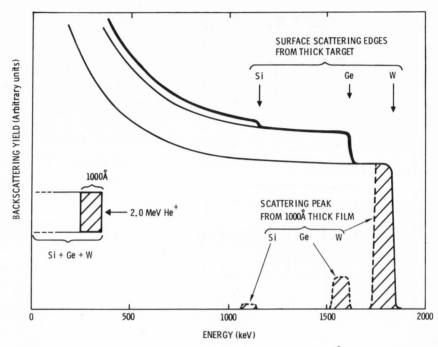

Figure 6. Hypothetical backscattering spectrum of a film (~1000 Å thick) and a thick target, both composed of equal numbers of Si, Ge, and W atoms. The shaded peaks indicate the spectrum from the film and the heavy line shows the spectrum from the thick target. The thin lines show the extrapolation of the W and Ge scattering yields of the thick target to lower energies.

in this energy range, and the spectra generally show increasing yields with decreasing energy.

In order to reduce the data from multiatomic targets to obtain compositional information, we must extract the scattering yield from each species at the depth we wish to determine the composition. For the data in Figure 6, this is essentially just the shaded peaks if we are interested in the first 1000 Å of the target. Note that for the thick target case the scattering yield for W must be extrapolated to lower energies below the step due to Ge surface scattering to obtain the scattering yield for Ge and similarly for W and Ge scattering at the Si edge. Once we have extracted the yields (Y_i) of all the constituents at a given depth, these are converted to atomic fractions by noting that

$$Y_i \propto \frac{N_i[d\,\sigma/d\,\Omega]_i}{[S]_i} \tag{6}$$

N_i and $[d\sigma/d\Omega]_i$ are the atomic density and the Rutherford cross section of the ith component of the target, respectively, and $[S]_i$ is the stopping power factor of the composite target for scattering from atoms of the ith component

$$[S]_i = \sum_j [S_j]f_j = \sum_j f_j[k_i\epsilon_j(\bar{E}_{in}) + \sec\theta\cdot\epsilon_j(\bar{E}_{out})_i] \tag{7}$$

where the f_j's are the atomic fractions of each constituent.

In many cases, the $[S]_i$ differ only slightly ($<5\%$), so that we can write for the ratio of the yields of two species

$$\frac{Y_A}{Y_B} = \frac{N_A}{N_B}\frac{[d\sigma/d\Omega]_A}{[d\sigma/d\Omega]_B} \tag{8}$$

Since the yields are evaluated at the same depth, the incident energies are the same for collisions with A and B type atoms and neglecting terms of order $(M_1/M_2)^2$, we have from equation (3)

$$\frac{Y_A}{Y_B} = \frac{N_A}{N_B}\left(\frac{Z_A}{Z_B}\right)^2 \tag{9}$$

Thus, we can evaluate the relative atomic concentrations of A and B by measuring the scattering yields from A and B. This method is absolute as long as the elastic scattering cross sections are known. Equation (9) has assumed the Rutherford cross sections which are valid in most cases. The final compositional fractions depend only on having measured scattering yields from all the atomic constituents.

As an illustrative example of the use of energetic ion backscattering to analyze the composition of a sample, we show in Figure 7, two spectra[13] from CoMo catalyst material supported on Al_2O_3 pellets. In the top part of Figure 7, is the spectrum from a new catalyst, and in the bottom half of the figure is the spectrum from a catalyst pellet which had been used in an experimental coal liquefaction reactor. The differences are obvious; Ti and S as well as large amounts of carbon are detected in the used material but were not present in the new material. In addition, the Mo yield is drastically reduced in the used material and the Co yield is increased over what was seen in the new material. This last observation, however, serves to point up a limitation of backscattering. Analyses using Auger spectroscopy and electron microprobe techniques indicate the presence of iron in the used material,[14] and in this case our measurements are not selective enough to distinguish between cobalt and iron.

From the data of Figure 7, it appears that there is no carbon in the

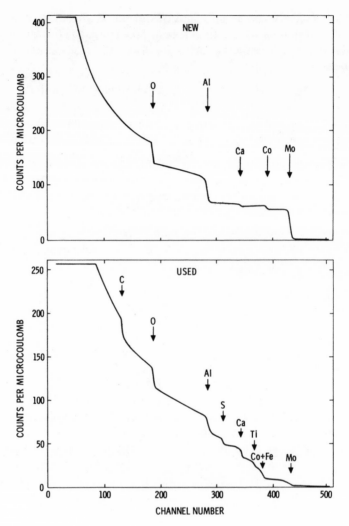

Figure 7. Ion backscattering spectra of new (top) and used (bottom) coal liquefaction catalysts. The spectra were both obtained using 2.0 MeV ^{4}He ions.

new catalyst, and it can be seen that the scattering yield from carbon in the used catalyst is difficult to determine accurately because of the rapidly changing scattering yield of heavier constituents. Analyses of the same samples using protons enable a determination of the carbon yield with greater accuracy because: (1) the carbon edge occurs at a higher fraction of the incident energy, (2) the scattering yield is changing less rapidly as a function of energy, and (3) the scattering cross section for 1.4-MeV protons from carbon-12 nuclei is about 6.7 times the Rutherford cross section[15] due to penetration of the carbon nucleus by the proton. Thus, proton-scattering experiments were used to determine the carbon present in these samples. The final compositions of the catalysts, as determined by energetic ion backscattering analysis, are shown in Table 1.

The previous examples have been for targets with no variation of composition with depth. However, one of the most powerful features of ion backscattering is the ability to nondestructively measure a depth profile of atomic composition. Since ion backscattering is sensitive to only the first few microns (for 2.0 MeV ^{4}He) of a target, many of the systems studied to date have involved thin films. The interactions of thin films with substrate materials and the interactions of multilayer thin-film structures are two areas of high technological relevance; in particular, many of these systems are used in the microelectronics industry. The solid–solid chemical reactions taking place between two thin films or between a thin film and its substrate are often ideally suited to investigation using ion backscattering. In fact, the interaction of thin metal films with silicon surfaces has been one of the most active areas in ion backscattering research.[4,5]

The system of thin-film sputtered tungsten on silicon–germanium alloy has been studied by Borders and Sweet.[16] This system is important because thin-film sputtered tungsten is used as the electrode on the hot end of radioactively heated SiGe thermoelectric generators. This electrode must operate at an ambient temperature of over 400°C for years without significant degradation.

In Figure 8 is shown an ion backscattering spectrum of a sample consisting of a 1200-Å layer of sputtered tungsten on bulk SiGe. The dots are actual data points, and the heavy line is drawn through the data points

Table 1. Composition of Coal Liquefaction Catalysts

Element	New pellet (At. %)	Used pellet (At. %)
Mo	3.6	0.5
Co & Fe	1.1	1.4
Ti	0	2.4
Ca	1.2	3.2
S	0	4.5
Al	23.5	11.3
O	56.5	47.3
C	14.0	29.4

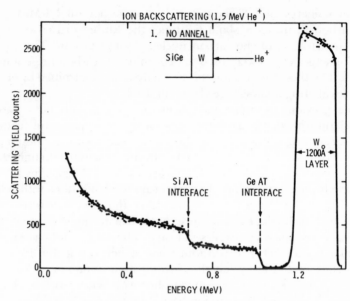

Figure 8. Ion backscattering spectrum of an as-prepared sample consisting of 1200 Å of sputtered tungsten on bulk $Si_{0.8}Ge_{0.2}$.

to guide the eye. Note that this spectrum was taken using 1.5 MeV He ions. The Si-to-Ge atomic ratio of 80 : 20 atomic percent was chosen to optimize the thermoelectric properties of the SiGe alloy.

The large peak in the scattering yield extending from about 1.4 MeV to just below 1.2 MeV is due to the sputtered tungsten film. The energy width of this peak, as measured between the half-height points of the high- and low-energy edges, respectively, can be used to obtain the thickness of the tungsten film by use of equation (5), as discussed previously.

The two scattering edges from Si and from Ge at the interface between the alloy and the tungsten film are at lower energies than if the Si and Ge were found at the surface. This shift in energy is approximately equal to the width of the tungsten peak and is due to the fact that ions must lose energy traversing the tungsten film before and after they collide with a silicon or germanium atom.

In order to simulate aging of the electrode at elevated temperatures, samples of sputtered tungsten on SiGe alloys were vacuum annealed at various temperatures. An ion backscattering spectrum from a sample annealed at 675°C for 20 h is shown in Figure 9. These results typify the behavior of many of the aged samples. The step or notch in the low-energy side of the tungsten scattering peak is due to a layer of tungsten silicide which has formed due to migration of silicon atoms into the film. By measuring the energy width of the step, we can calculate the thickness of the silicide layer which in the case shown is about 800 Å. The stoichiometry can be calculated in two ways. First, the scattering yield due to tungsten atoms in the silicide layer (under the step) can be compared with the scattering yield of silicon atoms in the silicide layer (shown shaded in Figure 9), and use of equation (8) enables the silicon-to-tungsten ratio to be calculated. Secondly, it has been shown[16] that if we know what the impurity atom is, we can calculate the stoichiometry by comparing the scattering yield of a pure element with that of a compound. In this case, we know the impurity is silicon and we can compare the height of the pure tungsten with that of the tungsten step. By either method we arrive at the same stoichiometry, WSi_2.

The fact that the energy gap between the low-energy edge of the tungsten peak and the germanium scattering edge remained approximately constant after thermal annealing indicates that Ge has not migrated into the film. However, the shape of the Ge scattering yield has changed; after annealing, the yield due to Ge is highest at the interface and decreases as the depth into the alloy is increased. The reason for this behavior is the migration of Si out of the alloy. This leads to an increase of the Ge-to-Si atomic ratio near the interface and to the observed spectrum. This outdiffusion creates Kirkendall voids in the alloy near the interface. These voids

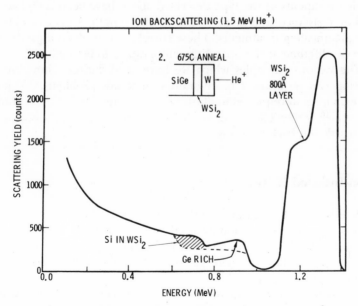

Figure 9. Ion backscattering spectrum of a sample consisting of 1200 Å of sputtered tungsten on bulk $Si_{0.8}Ge_{0.2}$ after vacuum annealing at 675°C for 20 h.

Table 2. Ion Backscattering

Measures
   1. Masses of target (absolute)
   2. Depth distribution of masses
Advantages
   1. Measurements easy
   2. Analysis easy
   3. Depth distribution nondestructive
   4. Sensitivity to heavy masses
Disadvantages
   1. Complex, expensive equipment
   2. Poor lateral resolution
   3. No elemental identification
   4. Low sensitivity to light masses

can be seen using a scanning electron microscope following electrode removal. Thus, a picture of bond failure due to thermal aging has emerged as due to the migration of Si into the tungsten film to form WSi_2, loss of Si in the alloy leading to void formation at the alloy–film interface, and a reduction in contact area as a result of void formation which causes increased contact resistance and decreased adhesion of the film to the semiconductor alloy.

Measurements of the type described above have been correlated with electrical resistance and mechanical bond strength measurements after identical annealing treatments. These correlations enabled a prediction of the service lifetime at the operating temperature to be made.[17]

The various examples which we have used illustrate how ion backscattering can be applied to laboratory materials problems. In general, these examples were picked to illustrate specific features of the techniques. In summary, Table 2 lists some of the major advantages and disadvantages of energetic ion backscattering.

4. Ion-Induced X Rays

A second area of research in the general field of energetic ion analysis of solids is analysis using ion-induced X rays. Actually, this technique is quite similar to electron-induced X-ray analysis such as used with the electron microprobe, but there are several unique features which can be useful in certain situations.

Energetic ions incident on a solid target have a high probability of ejecting inner-shell electrons from target atoms, thus creating inner-shell vacancies. Subsequent to the vacancy formation, a radiative transition can take place whereby an outer-shell electron falls into the inner-shell vacancy and the excess energy is carried off by a characteristic X ray. This process is schematically illustrated in Figure 10. The X-ray energy is characteristic of the atom from which the X ray was emitted so that this technique is element-sensitive rather than mass-sensitive as was ion backscattering. This process of the creation of inner-shell vacancies and subsequent radiative decay is exactly the same process used in electron-excited X-ray analysis. The differences in the techniques are due to the differences between ions and electrons in other interactions with the target atoms. Since most ion-induced X-ray work is performed using proton beams, most of the discussion in this section will be about proton-induced X rays. Generally speaking, however, the advantages and disadvantages of proton-induced X rays, when compared to electron-induced X rays, carry over to the case of heavy-ion-induced X rays.

A major difference between ion and electron excitation of characteristic X rays is the difference in the ranges of the respective particles used for inner-shell vacancy creation. For example, the range of a 100-keV proton in silicon is about 1.2 μm, whereas the range of a 100-keV electron

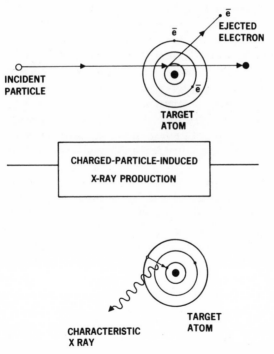

Figure 10. Schematic illustration showing the mechanism of production of characteristic X rays by charged particles.

is about 55 μm.[18] This means that for the same incident energy, the X rays excited by protons are, on the average, from atoms much nearer the sample surface. This increased sensitivity to the near-surface region can be very advantageous when investigating surface reactions. The sensitivity to the near-surface region is enhanced by the energy dependence of the X-ray production cross sections. For X-ray excitation using protons, the cross sections can be calculated[19,20] and reasonable agreement with experiment has been obtained. At low energies, the cross section can often be approximated as proportional to the fourth power of the incident proton energy.[20] This very strong energy dependence produces an X-ray yield which falls off very rapidly with depth, i.e., as the proton energy decreases. Another effect which must be considered is the absorption by atoms near the surface of X rays produced deeper within the target. This effect is also found in electron-excited X-ray analysis, but must be considered if quantitative measurements are being attempted. All of these effects tend to increase the sensitivity of proton-induced X-ray analysis to atoms near the surface and to decrease the sensitivity to atoms lying deeper within the target.

In theory, the depth dependence of the X-ray production, arising from the effects mentioned above, should allow measurement of the depth dependence of a species by measuring the X-ray yield from that species as a function of the incident beam energy. Reuter and Smith[21] have examined this problem and determined that the variation with energy of the kernel of the integral equation which must be solved is not sufficient to yield unique solutions. More recently, Pabst[22] has suggested that by varying the angle between the beam and the sample, depth profiles could be obtained. Feldman and co-workers[23,24] have used this method in studies of oxide films and Pt metallization interactions on GaAs and found that in some instances it can be a valuable complement to the information obtained using backscattering.

A second difference between ion and electron excitation of characteristic X rays is that ion-induced X rays give rise to much less bremsstrahlung radiation. This background radiation, due to acceleration of the ions and electrons, respectively, can cover up the characteristic X-ray signals from low-level impurities, particularly low-z impurities. The bremsstrahlung radiation results from the acceleration of the incident particles by the Coulomb fields of the target atoms, and intensity of bremsstrahlung radiation from an accelerated charge is proportional to the square of the acceleration. If we ignore relativistic effects, we find that the ratio of the intensities of electron-induced to ion-induced bremsstrahlung should vary inversely as the square of masses, i.e.,

$$\frac{I_{e^-}}{I_{p^+}} = \frac{M_p{}^2}{M_e} = (1836)^2 \simeq 3.4 \times 10^6 \tag{10}$$

In fact, it is observed that the low-energy background radiation for proton-induced X rays is significantly higher than would be predicted from the above analysis. This higher level arises from *knocked-on* electrons, which, although they have much lower energies, emit most of the bremsstrahlung observed in ion-induced X-ray analysis.

Most of the present work in proton-induced X rays was begun because of interest in the increased sensitivities which could be obtained because of the reduction in bremsstrahlung background radiation. The sensitivity of the ion-induced X-ray technique to low-level light impurities can be further increased by preparing thin foil samples. In this case, the number of target atoms available to accelerate recoil electrons is much reduced, and many of the recoils are ejected from the film in the forward scattering direction. In this way, very high detection sensitivities (i.e., 5×10^{11} atoms) have been obtained, and it may be possible to improve this to 10^7 atoms or less.[25] This has been shown to be particularly useful for the analysis of aerosol environmental samples. Cahill and co-workers[26] have used this method to analyze literally thousands of such samples.

Another advantage of ion-induced X-ray analysis over electron-induced X-ray analysis is the much straighter trajectories of the protons. Electron analysis beams tend to "bloom," that is, the lateral dimensions of the beams increase as the electrons penetrate the solid; this effect is much reduced using protons, for example. This feature has not been exploited, but in work where the lateral dimensions of the beam are critical, such as

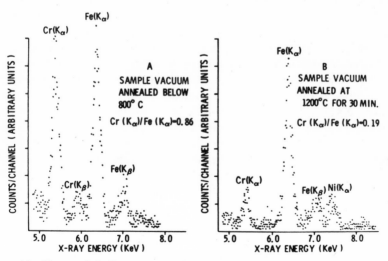

Figure 11. Characteristic X-ray spectra for a sample of 304L stainless steel (150 keV protons) after (A) vacuum annealing below 800°C, and (B) vacuum annealing at 1200°C. Reprinted with permission from "Surface Analysis Using Proton Beams," R. G. Musket and W. Bauer, in *Ion Beam Surface Layer Analysis*, J. W. Mayer and J. F. Ziegler (eds.), pp. 69–80, Elsevier Sequoia, Lausanne, 1974.

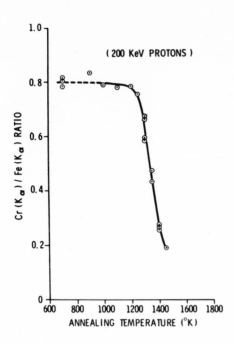

Figure 12. Annealing behavior of the Cr/Fe ratio as measured using 200 keV excitation of the K_α X rays. Reprinted with permission from "Surface Characterization of Stainless Steel Using Proton-Induced X Rays," R. G. Musket and W. Bauer, *Applied Physics Letters 20*, 411–413 (1972).

high-resolution microprobe work, protons may be preferable to electrons.

An interesting example of the use of ion-induced X rays in surface studies is illustrated by the studies of Musket and Bauer[27,28] on the surface composition of stainless steel following vacuum annealing. In particular, they used 150–400 keV proton-induced X rays to study polished specimens of stainless steel alloy 304L after 15-min isochronal vacuum anneals between 700 and 1450°K. Figure 11 shows a portion of the 200 keV proton-induced X-ray spectrum for a sample annealed to below (A) 1100°K and (B) to 1500°K for 30 min. Particularly obvious is the reduction in the yield of chromium K_α X rays at about 5.4 keV. Figure 12 shows the ratio of chromium to iron (K_α) X-ray yields as a function of incident proton energy. It was concluded that the chromium content in the surface region from 0 to 1 μm depth was depleted by 65%. This example indicates merely one use of proton-induced X rays to study solid-state reactions near the surface of a material. Because of the ability to identify elements and the very high impurity sensitivity which can be obtained, this analysis technique can be exceedingly useful. The equipment required is essentially the same as that for ion backscattering, and one of the most promising and least exploited areas is combined energetic ion backscattering and ion-induced X-ray studies.

Table 3. Ion-Induced X Rays

Measures: Elements in target (absolute)
Advantages
1. Low background
2. High sensitivity
3. Absolute elemental identification
Disadvantages
1. Complex, expensive equipment
2. Depth distribution difficult to obtain
3. No sensitivity to very light elements

In Table 3, we conclude this section by listing some of the major advantages and disadvantages of ion-induced X-ray analysis.

5. Nuclear Reaction Analysis

Nuclear reaction analysis is a general class of analysis techniques which depend on scattering events between the incident particles and the target nuclei. Ion backscattering analysis, discussed in Section 2, is a special case in which the scattering events are elastic. There are, however, many inelastic scattering events which can take place, and surface analysis, using these reactions, is usually called nuclear reaction analysis or nuclear microanalysis.

A schematic nuclear reaction is shown in Figure 13. A particle of mass M_1 is incident on a target atom of mass M_2 which is at rest. A reaction occurs between the two particles, and the products of the reaction (in this case, a light particle of mass M_3 and a heavier particle of mass M_4) are found to have energies E_3 and E_4, respectively. The sum of the energies E_3 and E_4 can be greater than E_1 (exothermic reaction), equal to E_1 (elastic scattering), or less than E_1 (endothermic reaction). The excess of the sum of the energies of the products over the energy of the incident projectile is called the "Q value" and can be either positive or negative. In general, the most useful reactions for nuclear microanalysis are those with large, positive Q values because the emitted products then have energies much higher than the particles which are elastically scattered.

In order for a nuclear reaction to occur, the incident particle must penetrate the nucleus of the target atom. This entails overcoming the Coulomb repulsion barrier between the incident ion and the target nucleus. The height of this barrier is proportional to the number of protons in the target nucleus and, thus, low-energy (0–2 MeV) incident ions can penetrate

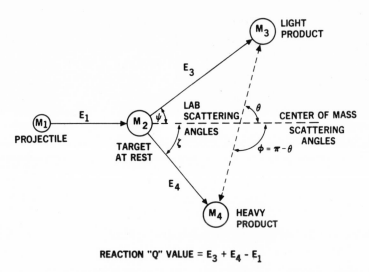

REACTION "Q" VALUE = $E_3 + E_4 - E_1$

Figure 13. Schematic illustration of the kinematics of a nuclear reaction.

the Coulomb barrier of low-z target nuclei but cannot penetrate the barrier of high-z nuclei. For this reason, nuclear reactions with low-z target nuclei are more likely than with higher-z nuclei for incident energies of 0–2 MeV. Similarly, reactions are more likely when using protons as the incident particles than with He, and so forth, because the Coulomb barrier is also proportional to the number of protons in the incident particle. For a 2-MeV proton beam, we can typically find reactions with target nuclei up to about Si. Deuterons are also very useful projectiles in this energy range for analyzing elements like C, N, and O because of the very low binding energy of the deuteron.

Because nuclear reaction analysis is most useful for light (low-z) target atoms, the technique is very complementary to energetic ion backscattering analysis. As was shown in Section 3, energetic ion backscattering is most sensitive to heavy target elements and relatively insensitive to light target elements, particularly when they are contained in a target along with heavier atoms. Thus, the high sensitivity of nuclear reaction analysis to a specific isotope of a light target atom, such as oxygen, can be used, while at the same time retaining high sensitivity to heavy impurities using backscattering.

The cross section for a given nuclear reaction is usually a very complicated function of energy. Some of the most useful reactions have cross sections which exhibit resonances or peaks where the magnitude of the

reaction yield rises to a much greater value than that found in the surrounding energy region. An example of such a reaction is the $^{18}O(p,\alpha)^{15}N$
reaction whose cross section at a laboratory scattering angle of 165° is
shown in Figure 14 for incident proton energies between 500 and 1000
keV.[29] The nomenclature $^{18}O(p,\alpha)^{15}N$ means that protons (p) are incident
on oxygen-18 target atoms and that the reaction yields nitrogen-15 and an
alpha particle (α). The Q value for this particular reaction is 3.98 MeV,
and the alpha particle has an energy of about 3.4 MeV if it is emitted at
a laboratory angle of 150°.

Note that protons which are elastically scattered from the ^{18}O nuclei
will have an energy of only 522 keV if 650-keV protons are used in the
incident beam. Also, the maximum energy for elastically scattered protons
is 650 keV regardless of the mass of the target atom.

Thus, a spectrum of the detected particles would show a peak from
the α particles at 3.4 MeV and a thick target backscattering yield of protons
at and below 522 keV. The yield of the α particles will be much smaller,
but since the proton yield does not interfere, the α-particle yield is easy to
determine.

Calculations of relationships between the Q value, the energy, the
laboratory scattering angle, and the masses of the participating particles
are straightforward. These kinematic equations are, however, quite lengthy
and the reader is referred to p. 141 of reference 10 for a convenient
compilation.

Table 4 lists some of the more useful nuclear reactions (with incident
energies <2.0 MeV) along with various pertinent data on each reaction.
The last column gives the thickness of Mylar film which may be placed
over the detector to absorb the incident particles which have been elastically
scattered but pass the higher-energy products of the nuclear reaction. This
detector shielding method is useful as there are generally many more
elastically scattered particles than nuclear reaction products because the
cross section for elastic scattering usually is larger than the nuclear reaction
cross section. In several cases shown in Table 4, there can be more than

Figure 14. Cross-reaction data for the
$^{18}O(p,\alpha)^{15}N$ reaction for incident proton energies of 500–1000 keV and a laboratory scattering angle of 165°. Reprinted with permission from "Microanalysis of the Stable Isotopes
of Oxygen by Means of Nuclear Reactions," G.
Amsel, *Analytical Chemistry 39*, 1689–1698
(1967). Copyright by the American Chemical
Society.

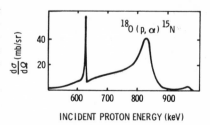

Table 4. Useful Nuclear Reactions for Surface Analysis

Nucleus	Reaction	Q value (MeV)	Incident energy (E_0) (MeV)	Emitted energy (MeV)	Approx. σ_{lab} (E_0) (mb/sr)	Mylar thickness (μm)
^{2}H	^{2}H(^{3}He,p)^{4}He	18.352	0.7	13.0	61	6
^{3}He	^{3}He(d,p)^{4}He	18.352	0.45	13.6	64	8
^{9}Be	^{9}Be(d,α)^{7}Li	7.153	0.6	4.1	1	6
^{12}C	^{12}C(d,p)^{13}C	2.722	1.20	3.1	35	16
^{13}C	^{13}C(d,p)^{14}C	5.951	0.64	5.8	0.4	6
^{14}N	^{14}N(d,α)^{12}C	13.574	1.5(α)	9.9(α)	0.6(α)	23
		9.146	1.2(α_1^0)	6.7(α_1^0)	1.3(α_1^0)	16
^{15}N	^{15}N(p,α)^{12}C	4.964	0.8	3.9	15	12
^{16}O	^{16}O(d,p)^{17}O	1.917	0.90(p$_1$)	2.4(p$_1$)	4(p$_1$)	12
^{18}O	^{18}O(p,α)^{15}N	3.980	0.730	3.4	15	11
^{19}F	^{19}F(p,α)^{16}O	8.114	1.25	6.9	0.5	25

one group of emitted particles. For example, the $^{14}N(d,\alpha)^{12}C$ can emit α particles with Q values of 13.6 MeV (α_0) or 9.1 MeV (α_1). For nuclear reaction analysis, careful thought should be given to the experimental set-up. Proper choice of the particular nuclear reaction to use, the incident beam energy, and the laboratory scattering angle can greatly simplify the data analysis and interpretation.

The group of G. Amsel in Paris has pioneered the use of nuclear reactions, particularly in studying electrochemical reactions. Much of their work has concerned the incorporation or movement of oxygen in thin films.[30] In this regard, they have made extensive use of the $^{18}O(p,\alpha)^{15}N$ reaction to analyze ^{18}O atoms introduced in controlled fashions. For example, they have studied the isotopic exchange of ^{18}O between oxides of tantalum or aluminum and aqueous solutions containing ^{18}O-enhanced or ^{18}O-depleted H_2O. They found that oxygen exchange is limited to the first few monolayers, indicating that oxygen self-diffusion in the oxides is very small.[31] Similar experiments have shown the existence of electric-field-induced isotopic exchange of oxygen in anodic oxide films on silicon.[32]

Some of the most interesting applications of nuclear reaction analysis have been the recent studies of hydrogen and helium in solids. There are many areas where these studies are relevant, and one of the most active current areas involves studying the interactions of energetic hydrogen and helium ions with materials which are being considered for use in the first walls of controlled thermonuclear fusion reactors (CTR).

A very useful reaction for study of hydrogen and helium in solids is $^3He(d,p)^4He$. When using this reaction, deuterium can be used to simulate hydrogen, or ^3He can be used to simulate ^4He. In some CTR studies, deuterium is in fact the hydrogen isotope of interest. Any nuclear reaction can be "turned around," that is, the target particle can be used as a projectile and the projectile as the target. However, if the original target particle is to be used as the projectile, and if it is much heavier than the original projectile, then the acceleration energy must be higher in order for the heavier particle to attain the same energy in the center-of-mass system. This is illustrated in Figure 15 which shows the cross section for the $^3He(d,p)^4He$ and $d(^3He,p)^4He$ reactions. These are the same reaction in the center of mass, but in the laboratory the first involves accelerating d to analyze for ^3He, and the second involves accelerating ^3He to analyze for d. The lower horizontal scale of Figure 15 gives the deuteron energy for analyzing ^3He, and the upper horizontal scale gives the ^3He energy for analyzing deuterium. It is easily seen that the incident energies scale with the projectile masses.[33]

Langley et al.[33] have discussed in detail the use of this reaction for studying hydrogen and helium isotopes in materials. They illustrate pro-

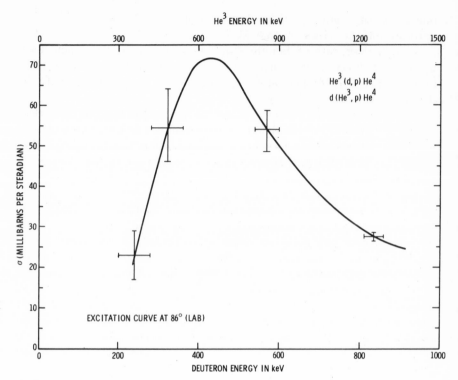

Figure 15. Cross-section data for the ³He(d,p)⁴He and d(³He,p)⁴He reactions at a scattering angle of 86° in the laboratory. Reprinted with permission from "Depth Distribution Profiling of Deuterium and ³He," R. A. Langley, S. T. Picraux, and F. L. Vook, *Journal of Nuclear Materials* 53, 257–361 (1974).

filing of deuterium in a multilayer target with the data shown in Fig. 16. The yield at energies below 1.0 MeV is due to backscattered ³He particles; the yield at energies above 1.5 MeV is due to the ⁴He reaction product of the d(³He,p)⁴He reaction. The two-peaked spectrum resulting from the multilayer target containing erbium deuteride films is shown. Because the resonance yield is quite broad, the two peaks cannot be completely separated, but further data reduction involving deconvolution of the reaction cross section is possible.

The use of reactions such as these for analyzing hydrogen and helium in materials has become an active area of research in the past few years. Most surface analysis techniques are not very sensitive to these elements, so nuclear reaction analysis provides a straightforward method for accurately measuring the amount and depth profile of hydrogen and helium present in the surface layer of a solid.

In Table 5 are listed several advantages and disadvantages, most of which have been previously discussed. However, a word needs to be said about the high fluences of analyzing particles which must normally be used in nuclear reaction analysis. The reason for these high fluences is that the reaction cross sections are generally rather small, and to obtain data which are statistically significant, many incident particles must be used. Again we emphasize that proper experimental design can be very important by picking the detector angle properly and maximizing the detector solid angle.

This technique is nondestructive in the sense that we do not remove layers of the target and repeat the measurement each time. The incident beam does damage the sample, however, and sputtering may also be a problem. To minimize problems such as these, it is generally advisable to use the minimum incident fluence compatible with the measurement under study.

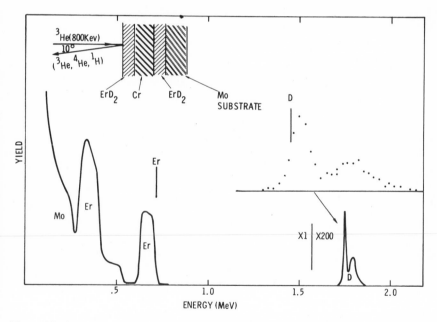

Figure 16. Spectrum of backscattered ^{3}He and ^{4}He from the reaction ^{3}He(d,p)^{4}He for a target consisting of a molybdenum substrate with thin films of ErD$_2$ and Cr as indicated in the upper left of the figure. Laboratory scattering angle is 170°. Reprinted with permission from "Depth Distribution Profiling of Deuterium and ^{3}He," R. A. Langley, S. T. Picraux, and F. L. Vook, *Journal of Nuclear Materials 53*, 257–361 (1974).

Table 5. Ion-Induced Nuclear Reactions

Measures
 1. Amount of specific isotopes in target (absolute)
 2. Depth distribution of isotope
Advantages
 1. Sensitive to only one isotope
 2. Depth distribution nondestructive
 3. Low background
Disadvantages
 1. Complex, expensive equipment
 2. Poor lateral resolution
 3. Low sensitivity to heavy masses
 4. Normally measure only one isotope at a time
 5. Large doses required normally

6. Comparison of Surface Analysis Techniques

The preceding sections have attempted to describe the various features of the three most widely used energetic ion-beam analysis techniques. In order to set the discussion in perspective, in this section we will briefly compare these energetic ion techniques with other modern surface analysis techniques, not with the intention of declaring one technique to be the best, but rather to illustrate how the various techniques complement each other. Where one method may be weak, another will be strong, and by combining the information obtained using different techniques, it is often possible to greatly increase our understanding of the problem.

In Table 6 are shown certain operating parameters or operational limits of various surface analysis techniques. These include backscattering (RBS), ion-induced X rays (IIX), nuclear reaction analysis (NRA), Auger electron spectroscopy (AES), low-energy ion scattering (IS), X-ray photoelectron spectroscopy (XPS), also known as ESCA, and secondary ion mass spectroscopy (SIMS). An important parameter which is not listed in the table is sensitivity. This is not included due to the fact that any comparison of sensitivity must take into account a large number of factors which could not be listed in a table. For example, RBS can detect ppm amounts of very heavy impurities in very light substrates, but would have a difficult time detecting 10 at.% oxygen on bulk uranium.

The discussions of this chapter have emphasized two points about energetic ion analysis techniques as compared to many other surface-layer analysis techniques. First, the results are absolute, because the interaction cross sections are known. Secondly, at least for the case of backscattering and nuclear reaction analysis, it is possible to obtain depth distribution information without destroying the sample. A third advantage of ion back-

Table 6. Surface Analysis Techniques

	RBS	IIX	NRA	AES	IS	XPS	SIMS
Excitation particle	Ion	Ion	Ion, Neutron	e^-	Ion	X Ray	Ion
Detected particle	Ion	X Ray	Ion, Gamma	e^-	Ion	e^-	Ion
Elemental detection	$Z \geq 1$	$Z \geq 4$	$Z \leq 16^a$	$Z \geq 3$	$Z \geq 2$	$Z \geq 1$	$Z \geq 1$
Elemental identification	Good	Excellent	Excellent	Excellent	Good	Excellent	Good
Depth probes (Å)	10^4	10^4	10^4	30	3	30	30
Lateral resolution	Poor	Poor	Poor	Excellent	Good	Poor	Good
Chemical information	No	No	No	Yes	Yes	Yes	Good
Standard sample required	No	No	No	Yes	Yes	Yes	Yes
Depth profile without sputtering	Yes	No	Yes	No	No	No	No
Complexity of equipment	High	High	High	Medium	Medium	High	High

^{a} Upper limit depends on maximum accelerator energy available.

scattering is that data interpretation is very easy and the measured physical effects can be estimated by visual examination of the data.

If we consider the techniques listed in Table 6 and ask what are the major disadvantages of each technique, it becomes clearer why combinations of various techniques are so advantageous. For backscattering, the most restrictive disadvantages are the lack of lateral resolution and the lack of absolute elemental identification which are provided by techniques such as scanning Auger spectroscopy or scanning electron microscopy with energy-dispersive X-ray analysis. Ion-induced X-ray analysis provides sensitive elemental identification but does not provide a nondestructive depth-profiling capability which backscattering can provide. Nuclear reaction analysis lacks sensitivity to heavy elements but can be combined with numerous other techniques to circumvent this disadvantage. All of the electron spectroscopic techniques need "standard" calibration samples in order to provide quantitative analyses; backscattering and nuclear reaction analysis do not.

Experimental interactions in which one technique can supply all the desired answers usually turn out to be quite rare. Access to a variety of methods of surface analysis and an understanding of the basics of each technique and the type of information it can supply are invaluable aids in the solution to many modern surface analysis problems. This review has provided information on the three major methods of energetic ion analysis and, it is hoped, has provided the reader with an interest in the techniques and the types of information they can provide.

ACKNOWLEDGMENTS

The author would like to acknowledge Dru Evans and Phil Kane as the organizers of the American Chemical Society Symposium on Modern Methods of Surface Analysis held at Arizona State University in Tempe, Arizona, June 3–5, 1976. This chapter is an outgrowth of a lecture presented at that symposium. Thanks are due to P. H. Holloway and J. Harris of Sandia Laboratories for a critical reading of the manuscript prior to its completion.

This work was supported by the United States Energy Research and Development Administration (ERDA) under Contract AT(29-1)789.

7. References

1. J. W. Mayer and J. F. Ziegler (eds.), *Ion Beam Surface Layer Analysis*, Elsevier Sequoia S.A., Lausanne (1974).
2. O. Meyer, G. Linker, and F. Käppeler (eds.), *Ion Beam Surface Layer Analysis*, Plenum Press, New York, 1976 (two volumes).

3. E. A. Wolicki, J. W. Butler, and P. A. Treado (eds.), *Ion Beam Analysis,* North-Holland, Amsterdam (1978).
4. J. A. Borders and S. T. Picraux, Characterization of silicon metallization systems using energetic ion backscattering, *Proc. IEEE, 62,* 1224–1231 (1974).
5. T. M. Buck and J. M. Poate, Ion scattering for analysis of surfaces and surface layers, *J. Vac. Sci., 11,* 289–296 (1974).
6. M.-A. Nicolet and W. K. Chu, Backscattering spectrometry, *Am. Lab.,* 7(3), 22–34 (1975).
7. D. V. Morgan (ed.), *Channeling,* John Wiley & Sons, London (1973).
8. D. S. Gemmell, Channeling and related effects in the motion of charged particles through crystals, *Rev. Mod. Phys., 46,* 129–228 (1974).
9. J. A. Cairns, C. L. Desborough, and D. F. Holloway, A new end-window variable geometry X-ray proportional counter, *Nucl. Instrum. Methods, 88,* 239–244 (1970).
10. J. A. Cookson, A. T. G. Ferguson, and F. D. Pilling, Proton microbeams, their production and use, *J. Radioanal. Chem., 12,* 39–52 (1972).
11. J. B. Marion and F. C. Young, *Nuclear Reaction Analysis,* p. 154, North-Holland, Amsterdam (1968).
12. J. F. Ziegler and W. K. Chu, Stopping cross sections and backscattering factors for ⁴He ions in matter, *At. Data Nucl. Data Tables, 13,* 463–489 (1974).
13. J. A. Borders, unpublished data.
14. P. H. Holloway, private communication.
15. H. L. Jackson, A. I. Galonsky, F. J. Eppling, R. W. Hill, E. Goldberg, and J. R. Cameron, The ¹²C(p,p)¹²C differential cross section, *Phys. Rev., 89,* 365–369 (1953).
16. J. A. Borders and J. N. Sweet, Ion-backscattering analysis of tungsten films on heavily doped SiGe, *J. Appl. Phys., 43,* 3803–3808 (1972).
17. J. N. Sweet and J. A. Borders, unpublished data.
18. Instruction Manual for Surface Barrier Detectors, Ortec, Inc., Oak Ridge, Tennessee.
19. F. Folkmann, Progress in the description of ion induced X-ray production; theory and implication for analysis, in: *Ion Beam Surface Layer Analysis* (O. Meyer, G. Linker, and F. Käppeler, eds.), Vol. 2, pp. 695–718, Plenum Press, New York (1976).
20. E. Merzbacher and H. W. Lewis, X ray production by heavy charged particles, in: *Encyclopedia of Physics* (S. Klügge, ed.) Vol. 34, p. 172, Springer Verlag, Berlin (1958).
21. F. W. Reuter III and H. P. Smith, Jr., Full range solutions for the measurement of thin-film surface densities with proton-excited X rays, *J. Appl. Phys., 43,* 4228–4232 (1972).
22. W. Pabst, Depth profile determination by ion-induced X-ray spectroscopy, *Nucl. Instrum. Methods, 120,* 543–545 (1974).
23. L. C. Feldman and P. J. Silverman, Depth profiling with ion induced X rays, in: *Ion Beam Surface Layer Analysis* (O. Meyer, G. Linker, and F. Käppeler, eds.), Vol. 2, pp. 735–745, Plenum Press, New York (1976).
24. L. C. Feldman, J. M. Poate, F. Ermanis, and B. Schwartz, The combined use of He backscattering and He-induced X rays in the study of anodically grown oxide films on GaAs, in: *Ion Beam Surface Layer Analysis* (J. W. Mayer and J. F. Ziegler, eds.), pp. 81–89, Elsevier Sequoia, Lausanne (1974).
25. T. B. Johansson, R. Akselsson and S. A. E. Johansson, X-ray analysis: Elemental trace analysis at the 10⁻¹² g level, *Nucl. Instrum. Methods, 84,* 141–143 (1970).
26. T. A. Cahill, Ion-excited X-ray analysis of environmental samples, in: *New Uses of Ion Accelerators* (J. F. Ziegler, ed.), pp. 1–71, Plenum Press, New York (1975).
27. R. G. Musket and W. Bauer, Surface characterization of stainless steel using proton-induced X rays, *Appl. Phys. Lett., 20,* 411–413 (1972).
28. R. G. Musket and W. Bauer, Surface analysis using proton beams, in: *Ion Beam Surface Layer Analysis* (J. W. Mayer and J. F. Ziegler, eds.), pp. 69–80, Elsevier Sequoia, Lausanne (1974).

29. G. Amsel and D. Samuel, Microanalysis of the stable isotopes of oxygen by means of nuclear reactions, *Anal. Chem.*, *39*, 1689–1698 (1967).
30. G. Amsel, Nuclear Microanalysis, in: *Physics of Electrolytes* (J. Hladik, ed.) Vol. 1, pp. 127–149, Academic Press, London (1972).
31. J. Siejka, J. P. Nadai, and G. Amsel, A study of oxygen growth laws of anodic oxide films of aluminum and tantalum using nuclear microanalysis of ^{16}O and ^{18}O, *J. Electrochem. Soc.*, *118*, 727–737 (1971).
32. M. Croset, S. Rigo, and G. Amsel, in: *Proceedings of the International Conference on MIS Structures* (J. Borle, ed.), pp. 259–268, Cent. Etude, Nucl. Grenoble–L.E.T.I. (1969).
33. R. A. Langley, S. T. Picraux, and F. L. Vook, Depth distribution profiling of deuterium and ^{3}He, *J. Nucl. Mater.*, *53*, 257–261 (1974).

3

A Synergic Approach to Graduate Research in Spectroscopy and Spectrochemical Analysis

John P. Walters

1. Introduction

Current research problems in analytical and physical chemistry are intricate, fundamental, complex, and diverse. Concurrent applications are perhaps even more so, covering such broad and elusive areas as metal alloy and trace analysis, energy sources and their byproducts, environmental pollution, and all-component human analysis in medicine. The scope of the research and application problems is so great as to challenge the intellectual capacity of the most ambitious. At the same time, the resources available to a single person to participate in and contribute specifically to solutions to these problems are limited.

Research in industry is, however, active in all the above areas and in others that are even less well defined. In part, this reflects the use of teams of selected specialists in the attack of multivariable research and application problems. Often this "team approach" is necessitated by the sheer complexity of the problem at hand (or the unforgiving and short deadline to be met). Alternately, the team approach may be elected to provide refreshing variety in the process of problem-solving and innovation in the solutions. It may occur through limited resources as well. Independent of the cause, the approach is well known[1] and has produced some remarkable products[2] and solutions to singularly difficult problems.[3]

John P. Walters • Department of Chemistry, University of Wisconsin, Madison, Wisconsin 53706

Research in the university environment may also involve a team approach. Teaching, when considered a research area in its own right, can be developed this way and may be quite effective when so developed or practiced.[4] The collaborative process involved in research and problem-solving is an appealing feature of any team approach and has been cited recently as a need in graduate education in analytical chemistry.[5]

However, the graduate situation in the university differs from its industrial counterpart in three essential ways. First, graduate students become educated through independent growth *in* their individual situations—they are not "trained" *by* them. The growth, by university definition, is an individual process. Second, graduate research often is exploratory and open-ended. Problems may be defined (and areas opened) rather than solved and finished. Third, a graduate student is learning by accumulating both positive and negative research results. Early judgement on the results of an experiment may diminish the quality of the "product," which is not just the solution to the problem at hand, but rather also is the effectiveness of the person involved in solving future problems that are, as yet, undefined.

Given the above, it is not clear that a literal transplant of team approaches from other environs into graduate research and education is appropriate. The approach introduced here is not such a transplant, but rather is traceable to the educational principles of learning through synthesis[6] and what is termed here as "continuity." The approach and the instruments involved allow the option for team collaboration between graduate students, if and when desired, but, at the same time, do not encourage or presuppose diminution of intellectual independence by requiring it.

When interactions and/or collaborations do occur, it is determined by the nature of the research progress on an individual problem. Interactions may occur either student-to-student, instrument-to-instrument, or past student and past instrument to present student and present instrument. The instruments are sufficiently flexible and stable that the timing is not of major concern.

Either combined or correlated action by different research participants may occur at any time in the investigation of complex experimental phenomena, even if one or more of the participants are no longer immediately present. The results of the action are not predictable by considering just the independent efforts of the individual participants, persons or instruments. The approach thus is designed to be synergic.

1.1. Guideline Principles

There have been four guideline principles active in the development of this synergic approach.

1. Realizing that the field of analytical chemistry is, as are other areas, moving faster and encompassing even more technologies than it has in the past, the approach was directed toward avoiding technological obsolescence.

2. Accompanying the above, it was sought to maximize the growth of the intellectual component of the research. It was not desirable that current graduate students repeat the same learning tasks that earlier students completed, unless such repetition was individually selected as one of a variety of practical alternatives in achieving growth.

3. It was sought to allow each generation of graduate students the option of pushing deeper into a core problem, opened in previous generations, or of opening new problems in a manner allowing subsequent generations the same options.

4. The approach, apparatus, and experiments were designed to accent creativity, enthusiasm, variety, and novelty; to avoid boredom; and to approach professional standards of quality. Avoided were those aspects of research suspected to eventually lead to mediocrity.

1.2. Overall Goals

These guideline principles were directed toward overall goals, both for the students and the professor. Sought for the students were:

1. Intellectual stimulation through participation in multivariable research problems of broad scope.
2. Intellectual freedom by emphasizing observation of what did happen in the research, rather than focusing attention on what should have happened.
3. Avoidance of restrictions on the imagination and creative ability of novice students that often result from restricted hardware or marginal experiment configurations.
4. The opportunity to develop creativity through the use of instrumentation that would concurrently encourage both precision measurement and exploration.

Sought for the professor were:

1. Intellectual growth through direct participation in the solution to fundamental problems at times judged to be opportune.
2. Consequential contribution to the solutions of existing or predicted analytical problems through knowledgeable communication and application of the results of fundamental research.
3. Opportunity for variety in daily activities by participation in local information exchange on diverse and changing research activities.

4. Personal professional satisfaction through credible endorsement of precision measurement over long times as an effective research tool.

1.3. Implementation

These guideline principles and overall goals have been implemented in specific research projects in the discipline of atomic emission and absorption spectroscopy and spectrochemical analysis. First work has been directed toward the development of methods and methodology for the multielement analysis of metal alloys. The developments have followed from research on the physical/chemical phenomena responsible for (or accompanying) the emission, absorption, or refraction of light. In studying the process of spark discharge, the directions taken followed the premise that insights gained from fundamental work would lead to applications, if the need for such applications was presented as an integral part of the work. The reverse premise was not used.

Atomic spectroscopy and spectrochemical analysis are appropriate vehicles for carrying the research approach to a communicable standing. They also are scientific disciplines in need of study, but of such intricacy as to warrant fundamental physical/chemical research. The latter viewpoint initiated this work, in particular with regard to developing an understanding of the process of spark discharge. The former viewpoint grew in importance as the work matured to the form communicated here. At present, the two are inseparably active.

1.4. Continuity

The need for a synergic aspect to the research accompanied an awareness of the experimental difficulty in acquiring enough experience with, and knowledge about, the process of spark discharge. Much knowledge is needed to improve performance in a spectrochemical analysis where both sampling and excitation are caused by spark discharge. It was apparent from prior research[7] and first experiments[8,9] that there were many phenomena active in controlling radiation emitted during and after[10] spark discharge. It also was apparent that detection and sorting of these phenomena would prove as challenging as their eventual control in an improved practical application.[11]

However, an approach to this dilemma was possible if two conditions could be met, essentially at the same time:

1. The creative efforts, and ideas, of several people would need to be translated into quantitative and permanent data and integrated between

present and/or future groups (generations) of such people, yet without diminishing the spontaneity and freedom that can lead to discovery.

2. The design ideas and construction efforts of groups of people of different talents and backgrounds in the fabrication of instrumentation would need to be available for both study and use by future groups, largely independent of the time between fabrication and use.

Given these two conditions, it was projected that a combination of discovery-oriented ideas and research-grade instrumentation would evolve. The combination, if made stable with time, would be available for use, independent of time. If the ideas and instruments were only iteratively modified during each use, the combination then could function as a new and flexible form of "instrument" to unfold research problems as intricate as those associated with spark discharge. Time, as such, would not be the limiting factor. Rather, the research would progress at a rate determined by the amount of synthesis possible.[12] By synthesis of the ideas and observations with the instrumentation of several groups of people, students and professionals alike, the desired synergic approach to improved spectrochemical analysis using complex discharges for sampling and excitation was projected to be both practical and exciting.

The word "continuity" has been used here to describe the above time-independent synthesis. Research accomplished with continuity between all participants, people and/or instruments, is the key to the synergic approach.

1.5. The Instrumental Vehicle

Early in the development of these ideas, it was apparent that a special type of spectroscopic instrumentation would be required if all of the overall goals were to be achieved. The goals would remain concepts unless time-independent synthesis between as yet unknown results and equally unknown new problems could be allowed, but not forced.

It was envisioned that in some cases an entire laboratory collection of individual instruments could be called upon to function for a period of time as a single, integrated instrument. At other times, individual instruments would have little or no interaction. The situation would follow individual researches and perspectives. It could neither be predicted nor scheduled.

To allow independent growth, spontaneous interaction, and continuity, the instrumentation would have to be, above all else, stable over unpredictable lengths of time.

For these reasons, the synergic approach was carried from an idealistic to a practical philosophy by an instrumental vehicle. Description of the instrumentation involved in this transport thus is an appropriate way to communicate the philosophy in explicit terms. The remainder of this chapter is largely so structured.

In retrospect, it is interesting that the first practical use of the approach of solving complex problems by synthesis, with interaction between participants, was not to execute spectroscopic observation of spark discharge. Rather, the same approach was used to build the special instruments that were envisioned as necessary to make such an approach practical. Communication of the idealistic philosophy in terms of these practical instruments thus provides the added benefit of presenting a partial report on the progress of the approach.

2. Changeable Instruments

Most of the perspectives taken in the design of the instruments and components presented here are hardware parallels to the overall goals listed in Section 1.2. Hardware designs thus are not at odds with these goals. To achieve this situation, some instrument design guidelines (equivalent to the guideline principles in Section 1.1) were followed from the start.

A first concern in preparing instrumentation for research with continuity was the effect such instruments could have on exploratory research. Any hardware developed could have a negative effect if it did not adapt with reasonable ease to a variety of exploratory projects. Further, the evolution of new instrumental arrangements could be restricted by the need to incorporate a previously beneficial component into the exploration, unless this too could occur simply.

The first perspective taken in designing the instrumentation thus was to facilitate exploratory research. Concurrent avoidance of experimental obsolescence followed as a consequence.

In instrumental terms, exploratory research in spectroscopy may become complicated by the number and nature of the components involved, even in first-approximation experiments. If the research is exploratory, the object of the search is largely unknown. Ideally, the instruments used must be sufficiently flexible to explore everywhere. Yet, they also must be sufficiently simple to allow reassembly into a prior configuration. If continuity is to accompany exploration, the instruments must allow a review of an earlier observation from an evolving perspective.

It is impractical, and contrary to one of the overall goals of this work, to require that very much new hardware be fabricated for each new exploration or observation. Exploratory work alone then predicates a flexible type of instrument. The instrument must be flexible, by being changeable. Its components must be interchangeable. The interchange must occur between both existing instruments, and those that have existed in past experiments or are expected to exist in future ones. Entire instruments themselves thus must be both changeable and interchangeable, but without extensive reconstruction. Such instrumentation may be approached if the appropriate design guidelines are observed.

2.1. Instrument Design Guidelines

Changeability and interchangeability are both achievable in an instrument until they are blended with the requirement that the instrument simultaneously allow continuity between quantitative experiments. Then an apparent contradiction between permanence and interchangeability surfaces. It is not immediately clear that an instrument can be adequately permanent during the execution of a precise, quantitative spectroscopic experiment, and, at the same time, be composed of components that are in fact interchangeable.

This apparent contradiction does not apply to all of the components in a spectroscopic instrument that is used for both precise measurement and exploration. For example, it is mainly an annoyance, if that, for many of the electronic components. This is evidenced in the effectiveness and versatility of the "bus" and "motherboard" features in MSI and LSI digital devices. Permanence and interchangeability conflict more in the traditional optical and mechanical components of spectroscopic instruments. In fact, in this area, they may become mutually contradictory for all but the simpler instruments, where exploratory work can be of token significance. For this reason, the first major design efforts were focused on the optical/mechanical aspects of the instrumentation.

2.1.1. Room to Work

In the above context, three design guidelines were selected to appear in all of the instrumentation. As an overriding concern, there first must be room to work, both in and around the components of the instrument. Interchangeability is unnecessarily complicated by not having enough, if not more than enough, physical room to place, test, and arrange components that will collectively form an instrument for some length of time. While, in general, this may appear somewhat trivial, it is all too often ignored as a continuous design guideline and is recalled only after a collection of components has settled into a form that becomes awkward by default.

2.1.2. Weight

As a subsequent co-guideline, weight, as such, is not considered an independent variable in either component or subassembly design. Unfortunately, it is easy to overreact to this guideline and to assume that weight is of little consequence. Instead of this, weight is considered as a result of another guideline, namely rigidity. Components then weigh whatever is appropriate for the amount of rigidity required to bring interchangeability and permanence into harmony. If a component becomes heavy, its weight becomes a tool to aid permanence, for which inertia is the major liability. Other separate tools then may be designed to facilitate interchangeability (e.g., carts, small cranes, etc.).

2.1.3. Rigidity

Rigidity is the most important guideline incorporated into the design of the instruments and components presented here. The concept of rigidity is deceptively simple. In a whole instrument, for example, rigidity does not imply that there be no vibration. Rather, it implies that all of the components composing the instrument vibrate in phase and to the same degree.

From the variety of ways that mechanical rigidity can be achieved, we have elected those taught by John Strong[13] and have adopted the "A-frame" (or gusset) as a primary design form. Variations experimented with using the A-frame as part of, for example, supports for modest-size components that are rigid in three demensions are diagramed in Figure 1. The "lollipop" is illustrated only for perspective. It is one form that we have learned to carefully avoid. The forms in Figure 1 also provide tangible examples of how rigidity is achieved as a practical consequence of allowing "room-to-work" and "weight-is-not-an-independent-variable" to operate in early generalized designs.

2.2. Instrumental Systems

When applying the above design guidelines, a final conceptual step is needed to remove the apparent contradiction between permanence and interchangeability. This is to expand the idea of an instrument from its common form of an assembly or collection of independent components to one of an instrumental system. The expansion may be aided by adopting the perspective that an instrumental system is in fact different from even a functioning collection-of-components instrument by virtue of three basic properties: (1) In the system, each component is designed to deliberately interact with all other components. (2) The specific design of any component should reflect the presence and functional design of all other com-

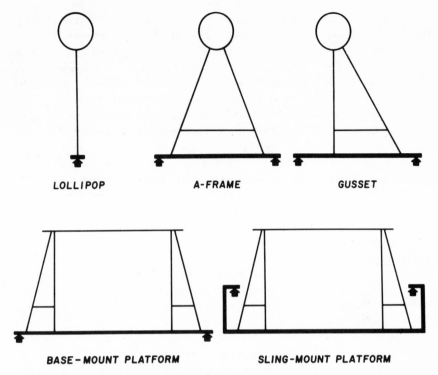

LOLLIPOP   A-FRAME   GUSSET

BASE-MOUNT PLATFORM   SLING-MOUNT PLATFORM

Figure 1. Basic A-frame structures used to achieve rigidity in mounting components and supporting platforms. The "lollipop" is neither an A-frame nor a rigid structure.

ponents. (3) Each component of the instrumental system is first referenced to a standard that will be common to all components, and then as a consequence referenced to the other components. It is the last property that is most important in the design of our instrumental systems, and, when applied to the optical and mechanical components, gives them their unique appearance. It also is due to this property that changeability within a system, and interchangeability between systems, is realized without loss of permanence during the execution of an experiment, and by which instrumental continuity is achieved.

2.2.1. Advantages

At the risk of overgeneralization, it is instructive to consider the advantages of adopting an instrumental system approach in favor of a collection of components, particularly for use by people who can overreact to what they perceive as a "failure." There are four identifiable advantages.

Component Balance. By designing the system for deliberate, as opposed to accidental, component interaction, the balance between the components is visible. This makes the system esthetically pleasing, and this in turn helps to offset the anxiety that can result from anticipation of an unknown interaction between components. A person can be more relaxed, and concurrently more innovative, with an instrumental system.

External Standardization. Then, since an instrumental system has an identifiable standard to which all components are first referenced, the standard may be located external to any specific functioning system. This suggests that as it is necessary to alter, replace, add to, or exchange components of any one system, the rough work of initial fitting and testing need not interfere with the precise work of quantitative measurement. This can encourage flexibility and inventiveness.

Obsolescence. The concept of instrumental obsolescence can be changed to one of growth and refinement. Since the components of an instrumental system are designed with each other in mind, their common characteristics can be saved during the time that the functions of the system are expanded or changed. First efforts can remain first efforts, not become obsolete failures. If the first efforts involve time in a machine shop, then this advantage is nontrivial.

Flexibility. Perhaps of most importance, and largely because an instrumental system has components that are interchangeable with other systems, the concept of a finished, bounded, special-function instrument can be replaced with one of a fixed foundation to hold a variety of components. The components may be arranged into one of a variety of functions at any one time.

2.2.2. Liability

It is clear, however, that the above four advantages are not realized without liability. To design and construct the common reference for all components requires a realistic overview of the probable uses of the instrumental system, as well as some insight into at least the types of components that will be needed for such uses. Thus, the research that the system will serve must have passed through at least one exploratory phase and the expenses attendant amortized over the long run. In fact, initial cost and labor associated with the development of an instrumental system are generally higher than required for other types of instruments, and must be invested with patience, without the luxury of direct return of experimental results. The work also may be demanding and potentially professionally dangerous. It is complicated by the knowledge that a fundamental mistake made during the early design and construction phases can be difficult to remedy at a later time.

3. Specific Hardware

3.1. Selection of a Common Standard

To translate the previous design guidelines into hardware amenable to research in spectroscopy first requires an assessment of what part of the area can serve as a common technical theme to allow design of the common standard for all components of whatever systems evolve. This is a nontrivial choice. However, spectroscopy and spectrochemical analysis can be viewed in three phases to facilitate the choice. They are: (1) those devices and experiments that involve the production or energetic alteration of light; (2) those that establish how light is transferred through the instrumental system; and (3) those that involve the detection of light and the subsequent feedback of processed information back to the production phase.

Viewing these three phases as a connected loop allows the selection of one as dominating the experiment by virtue of being the most difficult to control experimentally. If the common standard required for the instrumental system is then designed to facilitate experimental work in that phase, it is expected that the other two phases will become experimentally tractable in a wide variety of experiments.

3.1.1. Transfer of Light

Modern electronics is evolving at a bewildering pace,[14] and new types of thinking are resulting from laser research.[15] It was expected that neither the production nor the detection or processing of light posed problems of such fundamental magnitude that they should determine the properties of instrumental systems meeting the needs described by the word continuity. Rather, the transfer of light is the most difficult phase. Here, transfer is interpreted broadly, to include such "techniques" as sorting by wavelength and time and chemically passive or active scattering, along with the more traditional viewpoints of light shaping (i.e., collimation, focusing, etc.).

In applying the word "transfer," the concept is expanded from transfer to or transfer from a physical location to that of transfer through the instrumental system. The physical paths defined are viewed, for design purposes, without the restriction that they must start or stop at any specific location in the instrumental system. The components of the systems, as they are designed to account for each other by being first referenced to a common standard, then do so by being designed around the common denominator of facilitating the transfer of light through whatever system of which they are a part. The physical size of the system is not at issue.

3.2. FMR and ERL

The common standard to which all components are referenced (including light sources, dispersers, and detectors) is a mechanical reference plane, preferably, although not necessarily, at earth level. All of the instrumental systems presented here clearly show this choice in the form of a common, fixed mechanical reference (FMR). From the fixed mechanical reference, an experiment reference line (ERL) is generated to allow the system to perform some specific (but not necessarily fixed) function.

3.2.1. Pedestal, Rails, and Rider

The general ideas and their associated terminology used to prepare an instrumental system having a known FMR and ERL are diagramed in Figure 2 for one part of a general system. The FMR plane is defined, with respect to the building floor, to be at earth level. This is done by first suspending two centerless-ground and polished steel rounds (called "rails") from a custom "pedestal," typically with some form of V blocks.[16] One rail is set to describe a straight and level line, while the other is set only to define the reference plane as being at earth level. The plane is made available to define the ERL by contact mounting a "rider" of known and toleranced dimensions on the two rails via "V-and-flat" surfaces, i.e., in the semikinematic manner described by Strong.[13] Specific components of a variety of shapes and weights are positioned along the ERL. They are positionally referenced to the FMR plane by a variety of custom A-frame mechanical supports.

The steel round that is set to be straight and level is designated as the "reference rail." Although the top surface of the V-and-flat rider physically establishes the laboratory FMR plane, the top edge of the reference rail is, in many cases, used to locate the top of the rider. Most of the pedestals currently in use are largely open to the laboratory floor. When the ERL is set relative to the FMR, there then is substantial room around the ERL, including below it. This allows all of the A-frame component supports diagramed in Figure 1 to be achieved even for large components, e.g., a large monochromator.

3.2.2. Practical Devices

The components generalized in Figure 2 are shown in a first specific design (fabricated in 1969) in Figure 3. A functional description of the parts called out in Figure 3, and design dimensions for the rails and riders for all systems in current use, are listed in Table 1. In operation, the rail mounts are placed on the machined top surface (B) of the A-frame, cast-

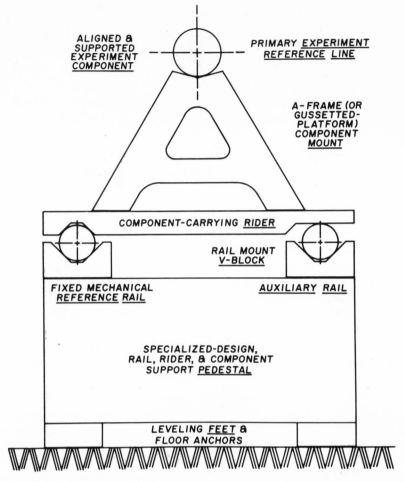

Figure 2. Diagram naming the basic parts used to establish an experiment reference line (ERL) that is dimensionally locatable from a fixed mechanical reference (FMR).

aluminum, support pedestal using the predrilled hole pattern (D). The rails lay in the jaws of the rail mounts. Their height and relative position are set by adjusting the separation of the jaws. The centerlines of the two rails are separated by a distance of 15 in. (381 mm).

3.3. An Experiment Bed

The pedestal and rails in Figure 3 do not, in isolation, constitute either the FMR or an instrumental system. They are only a part. To first establish

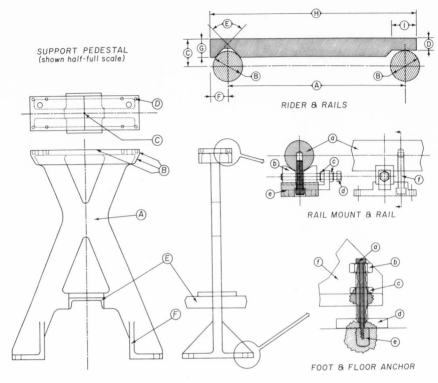

SUPPORT PEDESTAL
(shown half-full scale)

RIDER & RAILS

RAIL MOUNT & RAIL

FOOT & FLOOR ANCHOR

Figure 3. A specific example of a pedestal, rails, mounts, and a rider that are assembled to provide a working ERL and FMR. Dimensions for the rider and rails are given in Table 1. The dimensions are standard for all instrumentation described here (see also Figure 8).

the FMR, it is necessary to connect several pedestals together, into an assembly, and then to align the assembly. Once aligned, the FMR is set, and then the ERL may be defined.

For example, in Figure 4 is shown one type of pedestal assembly (commonly referred to as a bed). The support pedestals shown in Figure 3 are connected together, first with an aluminum "U" channel (part E in Figure 3) and then by the steel rounds themselves. The process of alignment accompanies the connecting. The assembly shown in Figure 4 was the first built. Since it was desired that the experience serve as a model for other assemblies, the parts involved and the alignment procedure used are characterized by greater attention to detail than is presently practiced. However, by discussing the alignment procedure as it was done, rather than as how it should be done, laboratory significance of the resulting FMR will be more evident.

3.3.1. Pedestal Alignment

The first step in connecting the assembly shown in Figure 4 was to locate the individual pedestals approximately 34 in. (864 mm) apart. Bending calculations indicated that a steel round of 2.5-in. (63.5-mm) diameter would deflect less than 0.005 in. (0.13 mm) under a 400-lb (174-kg) load if

Table 1. Detailed Parts of Figure 3

Item	Description
Support pedestal	
A	Cast aluminum A-frame support body
B	Machined mechanical reference surfaces
C	Drill point to locate center of top surface
D	Toleranced holes to locate rail mounts
E	Aluminum channel to interconnect pedestals
F	Pedestal foot to hold floor anchor
Rail mount and rail	
a	Centerless ground and polished steel rail
b	Sliding jaw to contact and support rail
c	Lock and adjusting nuts on jaw drive screw
d	Left and right hand threaded jaw drive screw
e	Notched base block to hold sliding jaws
f	Bolt to pull rail into line contact with jaws
Foot and floor anchor	
a	Bolt to hold support screw to support plate
b	Height-adjusting, threaded support screw
c	Lock nut on support screw
d	Support plate with conical seat for screw b
e	Lead insert in floot to hold bolt a
f	Cast aluminum foot to hold screw b
Rider and Rails	

	Dimension	
	Inches[a]	Millimeters
A	15 (standard)	381
B	2.500	63.50
C	2.250	57.15
D	1.000	25.40
E	90 degrees	100 grads
F	1.5 (nominal)	38
G	1.5 (nominal)	38
H	17.5 (nominal)	445
I	2.12 (nominal)	53.8

[a] Throughout this paper, all components were specified, toleranced, and fabricated using English dimensions and machine tools reading in English units. Metric dimensions are listed to comply with current practices, and are simply conversions from the English units using the factor 25.4001 millimeters/inch.

A

LOCATION	HEIGHT(in.)	HEIGHT(mm.)
a	0.0092	0.234
b	0.0118	0.300
c	0.0107	0.272
d	0.0117	0.297
e	0.0100	0.254
f	0.0112	0.284
Ave.	0.0108	0.274
S.D.	0.0010	0.026

B

LOCATION	HEIGHT(in.)	HEIGHT(mm.)
g	0.0089	0.226
h	0.0110	0.279
i	0.0102	0.259
j	0.0123	0.312
k	0.0110	0.279
l	0.0107	0.272
Ave.	0.0107	0.272
S.D.	0.0011	0.028

C

LOCATION	HEIGHT(in.)	HEIGHT(mm.)
A	0.0120	0.305
B	0.0132	0.335
C	0.0125	0.318
D	0.0124	0.315
E	0.0116	0.295
F	0.0120	0.305
Ave.	0.0122	0.312
S.D.	0.0005	0.014

D

LOCATION	WIDTH(in.)	WIDTH(mm.)
G	0.0112	0.285
H	0.0113	0.287
I	0.0110	0.279
J	0.0120	0.305
K	0.0125	0.318
L	0.0110	0.279
Ave.	0.0115	0.292
S.D.	0.0006	0.016

E

LOCATION	LEVEL(in.)	LEVEL(mm.)
M	0.0003	0.008
N	0.0006	0.015
O	0.0006	0.015
P	0.0003	0.008
Q	0.0003	0.008
R	0.0003	0.008

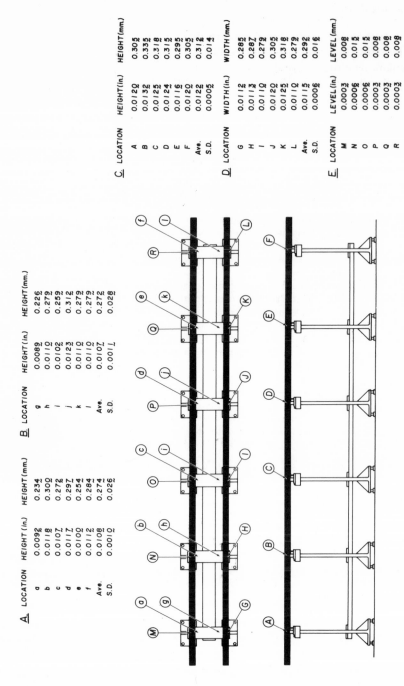

Figure 4. Assembly of multiple pedestals with long rails to be aligned into an experiment bed.

supported at two points 34 in. (864 mm) apart. The pedestals then were set in a straight line by laying a long, taut "piano" wire along the centerline scribed into the machined top surface of each pedestal (parts B and C in Figure 3). The top surface of each pedestal was then initially set to earth level using a machinist's spirit level, while at the same time the taut wire was used to maintain the top of each pedestal at approximately the same height. Then a Brunsen, model 659778, calibrated surveying telescope (itself carefully leveled) was positioned to sight down the assembly. A target was placed on either corner of the top of each pedestal to allow final adjustment to be made quantitatively. The adjusting bolts in the feet of each pedestal were interactively adjusted to raise or lower the target to a position within the calibrated range of the telescope. The aluminum channel connecting the pedestals was then bolted into place, and the telescope used to bring the top surfaces of all of the pedestals to a toleranced common height. The pedestals were then anchored to the floor as shown in Figure 3.

The final heights of the pedestals are shown in Tables A and B in Figure 4, for the two positions diagramed. By comparing the relative heights measured at positions (a) through (f) to each other, it may be seen that one side of the pedestal surfaces falls along a straight line to within a standard deviation of 0.0010 in. (0.026 mm). Locations (g) through (l) show a similar situation for the other side. By comparing the relative height measured at position (a) with that at (g), (b) with (l), etc., it is evident that the top surfaces of each pedestal are level to within a maximum deviation of 0.0008 in. (0.020 mm) over their 18-in. (457-mm) width, i.e., to within 0.002 degree.

3.3.2. Rail Alignment

Once the pedestals were positioned, the rail mounts were bolted to them, again using the taut wire to ensure that the jaws were properly centered and in a straight line. The rails were then placed in the mount jaws and brought into gentle contact with the jaw faces using the hold-down bolt (part f in Figure 3). The lower rail in Figure 4 was designated as the reference rail. Using the surveying telescope and a target positioned along the rail, the jaws of the mounts holding the reference rail were interactively adjusted to cause it to conform to a straight and level line.

For example, Table C in Figure 4 indicates that the height of the reference rail is constant, as measured at the six locations labeled A through F, to within a standard deviation of 0.0005 in. (0.014 mm). Table D in Figure 4 indicates that the same rail is straight, as measured at the six locations labeled G through L, to within a standard deviation of 0.0006 in. (0.016 mm). The auxiliary rail need have only its top edge coplanar with that of the reference rail. It was leveled using a machinist's level having a sensitivity of 0.0003 in. (0.008 mm) per 15-in. (381-mm) span. Table E. in Figure 4 indicates that, for measurement at the six locations M through R, the auxiliary rail is level to within the tolerance of the spirit level. This last step then defines the FMR as the top edge of the reference rail. The ERL is set relative to it.

4. Choosing and Locating the Experiment Reference Line

With a fixed mechanical reference available, an experiment reference line may be numerically established. The choice must reflect a number of variables. Once chosen, the same ERL will be established for all instrumental systems, implying that all components placed on and around the ERL will have their supports machined to its demensions. Monochromators, polychromators, spark and arc sources, mirrors and gratings, and all forms of auxiliary light sources and detectors are considered components of an instrumental system. It is clear that specifying an ERL should be done with foresight.

4.1 Primary ERL Location

Two guidelines were followed in setting the ERL. Figure 4 indicates that a support bed may be long. This may be a typical situation. When necessary to achieve width in an experimental system, other rails may be added as required, referenced to the FMR of the primary bed. Thus, there will be a primary ERL directed along the primary bed, and other secondary ERLs referenced to it. Considering the variety and probable sizes of the components to be placed both on and around the ERL, as well as the desire to design with the principle of transferring light through all of the components composing the experiment, it is appropriate to locate the primary ERL between and above the two rails composing the primary bed. The two guidelines followed were: first, to position the primary ERL such that there was ample room to work all around it (i.e., both above and below) and, second, to position it such that all A-frame members used to support components could be sized to maximize the rigidity of the supporting assembly.

4.1.1. Height

Having specified the center-to-center separation of the primary rails at 15 in. (381 mm), the horizontal dimension of the ERL was set at 7.500 in. (190.5 mm) inside of the reference rail centerline. The ideal vertical dimension of the ERL would then be 10.000 in. (254 mm), such that a 3:4:5 right triangle would be formed between the centerline of the reference rail, the ERL, and the vertical ERL axis, However, the centerline of the reference rail is not available as a direct measuring surface. Instead the top of the rail is used in the lab for vertical measurement, and the inside (3 o'clock) side used for horizontal measurement. Also, the rails will first hold a V and flat rider when supporting a component. Thus the ERL vertical dimension was set at 10.500 in. (266.7 mm) above the top surface of the reference rail, or 9.500 in. (241.3 mm) above the flat top surface of a supported rider. In this manner, the contact points of an A-frame gusset used to support a component directly on the ERL will describe a 3:4:5 triangle when the rider is contacted once at the center of the two rails and then at a point 3/8 inch (9.52 mm) inside the centerline of the reference rail. This is an adequate approximation to a true 3:4:5 triangle support.

4.1.2. Length

The above steps define the position of the primary ERL. However, placing a component on it, or at a known distance from it, presupposes that the ERL may be physically located at any point along the bed. Three techniques have been used to date to accomplish this, using apparatus such as shown in Figure 5. The most straightforward approach is to suspend a low-power helium–neon laser (part 1) with a kinematic, adjustable mount (parts 2 and 3) from a rider (part 4) contacting the fixed reference rail (part 5). The laser beam typically is aligned to the ERL with a variety of devices (4-quadrant photodiode, height guages, targets, telescopes, taut wires, etc.). It then is made one component of a particular system.

Figure 5. Laboratory tools routinely used to locate an ERL from an FMR anywhere along an experiment bed or between beds.

The center of the laser beam, or of a diffraction pattern thereof, is taken as the ERL. Although the center of the beam can be located only to a tolerance of ±0.1 mm (one standard deviation on 10 telescopic measurements at 6 m from the laser), it offers the advantage of defining the transfer of light completely through such components as monochromators and polychromators. Such large, sealed components have slits or similar devices that often act as stopping points for light transfer in a "collection-of-components" instrument.

The laser also offers the decided advantages of nondestructive contact, and of, literally, showing the ERL through all components of large instrumental systems. These last two points have proven so popular with those using the systems that it is now considered necessary to have one or more "alignment lasers" operating simultaneously with an experiment for each instrumental system. In the event that a laser vignettes or blocks another secondary ERL in use during an experiment, then the rider carrying it is removed from the rails, not disturbing the alignment of the laser. The rider may be placed back on the rails at any time, in the same location it originally occupied (±0.0005 in., 0.013 mm) by touching its edge against a rounded-end screw (part 7) in a "rail-stop" (part 6) collar that is clamped to the reference rail (part 5) prior to removal of the rider. The rail-stops are effective in aiding the "changeable and interchangeable" component concept both during and between experiments.

4.1.3. Absolute Measurement

For some experiments, it is necessary to place mechanical and/or optical components on or around the ERL with greater absolute numerical accuracy than allowed using the laser. In this case, the ERL is first located using crossed height gauges (parts 10 and 11 in Figure 5) mounted on a special "reference rider" (part 8) using rigid, flat, right-angle support plates (e.g., part 9) whose dimensions are accurately measured to ±0.0002 in. (0.005 mm). The component to be mounted is then supported in an assembly that has premachined reference surfaces. A machinist's transfer stand is used to mechanically transfer the position of the ERL, as defined by the junction of the pointers of the height guages, to the reference surfaces on the component supports (just as is done in setting up a milling machine). This operation is used with an accuracy of ±0.001 in. (0.025 mm), even for large components. Each of the present instrumental systems may have one or more reference riders with single or crossed height gauges as dedicated components. Rail-stops also are used to allow precise repositioning of a reference rider, again during or between experiments.

4.2. Transfer of the ERL

The above procedures allow a primary ERL to be located with acceptable accuracy for any one experiment bed. However, they do not address the point of continuity in research resulting from component interchangeability between other experiment beds. No fundamental limit is desirable on the pedestal style or number of such beds that will be established and used during the course of the research. Instead, it is stipulated that the same ERL be established on, and/or transferred between, any experiment bed at any time. This latter stipulation poses additional requirements on the whole instrumentation system, as follows:

1. A special, permanent experiment bed should be available to define the ERL. This bed will serve to transfer the ERL to all other beds. It should be largely dedicated to this function. It should be characterized by stability and should hold alignment over long periods of time (years) to a high degree. It then may function as a primary ERL standard.

2. All rails used in the fabrication of additional experiment beds should have the same concentricity and diameter. They should be aligned to tolerances similar to those listed in Figure 4.

3. A large quantity of riders should be available, all of uniform dimension and toleranced to conform to the dimensions in Figure 4.

4. As additional experiment beds are constructed, the base support pedestals may be of unique, specialized design for specific types of experiments, but each bed should have the same FMR (e.g., Figure 3). All ERLs should be at common dimensions relative to the FMR.

Each of the above points were felt to be sufficiently important to warrant design effort. For clarity, data relating to all four listed points are presented below.

4.2.1. Transfer Bed

Construction of a dedicated transfer bed was begun in the fall of 1972. The primary design guidelines were stability and permanence. Details of this transfer bed are shown in Figure 6. The two-rail nature of the bed is similar to that shown in Figure 4, except that the reference and auxiliary rails are permanently mounted on a solid concrete support pedestal (inset C, lower) that rests on pillars that are separated from the building floor. In contrast to Figure 3, the rail mounts are solid V blocks, machined to fixed dimensions, and are not adjustable (inset A). Six steel plates (part B) are fastened to the concrete pedestal (part A) with epoxy resin at the indicated rail-support locations along its length and held down by threaded studs (parts D and E). These plates each have the traditional V, cone, and

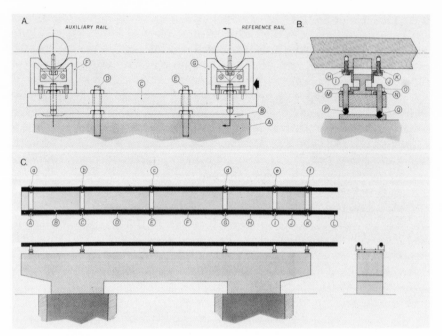

Figure 6. Components and their assembly to prepare a standard bed used to transfer a common ERL to other experiment beds.

flats cut into their surfaces for kinematic mounting (e.g., reference 13). However, provision is made for four contacts, rather than the usual three, to allow more even load distribution and stability.

The V blocks to hold the rails (parts F and G) are mounted on a thick, steel, adjusting plate (part C). This plate then is kinematically mounted to the mating plate on the support pedestal. The adjusting plate is moved to align both rails. Once adjusted, the assembly is locked into position using the two threaded studs (parts D and E) with large lock nuts.

Further stability and permanence are achieved with this bed by pulling the rails into firm contact with each V block by brackets and bolts (inset B, parts H and I) in addition to bolting the V blocks to the adjusting plate. The screws (parts L and O) and contact balls (parts P and Q) used to locate the adjusting plate on the pedestal also are locked into place (parts M and N).

Permanence was achieved for the transfer bed in the above manner, but two penalties were also accrued. First, the fixtures used to hold and align the rails are hard to adjust. The bed shown in Figure 4 was aligned in a matter of days. This one required careful and tedious adjustment over a period of weeks. Second, by locking the steel rails to a concrete base at six points, seasonal temperature variations can cause the rails to buckle due to the different coefficients of expansion of the two dissimilar materials. These two penalties were accommodated directly, i.e., by taking the necessary time to do the alignment and by adjusting the room air supply to allow appropriate temperature regulation.

Alignment of the transfer bed followed a procedure similar to that used for the bed shown in Figure 4. The reference rail was straightened by tapping the V blocks (black arrow in Figure 6, inset A) for horizontal adjustment, while interactively rocking the adjusting plate for vertical. When measured with a surveying telescope and target at the locations marked A through L in Figure 6, the centerline of the reference rail showed a maximum deviation from a true straight line not greater than ±0.001 in. (± 0.025 mm) in both horizontal and vertical directions. The auxiliary rail was brought into the same plane as the reference rail, as measured with a machinist's level at points a through f, to within the same maximum deviation. The center-to-center separation between the rails, as measured at points a through f, was adjusted to be 15.0 ± 0.001 in. (381 ± 0.025 mm).

The above tolerances establish a minimum requirement for the transfer bed if it is to function satisfactorily in holding the primary ERL. They have been checked several times during the past five years. In a recent test, the ERL was established to coincide with theoretical coordinates using a helium–neon laser, crossed height gauges, and a four-quadrant, position-sensitive photodiode to within an estimated tolerance of ±0.005 in. (±0.127 mm) absolute. Then, the FMR was measured, as indicated by the out-of-plane wobble of the top surface of a flat rider as it was moved along the rails. Deviations in the ERL from a true straight line were measured by sliding the four-quadrant photodiode along the rail and moving it back onto the original ERL with precision indicators. These data are shown in Figure 7. They are within acceptable limits for a five year index of the stability of the transfer bed.

4.2.2. Rails

As additional experiment beds were constructed, the ground and polished steel rounds used as rails were purchased at different times and from different material lots. Transfer of components between experiment beds, while maintaining the same primary ERL, requires that all the rails be equally round and of the same diameter. Again, the standard diameter of the rails will be that for the transfer bed (Figure 7). For this bed, 24 measurements were made of the rail diameters, approximately every meter on both the reference and auxiliary rails. The average diameter was measured to be 2.4986 in. (63.465 mm) with a standard deviation of 0.0002 in.

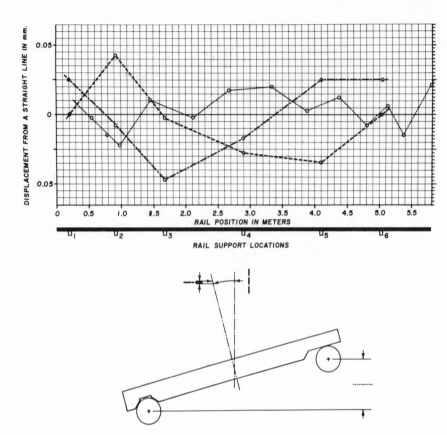

Figure 7. Dimensional, point-check measurements made to verify the stability of the transfer bed shown in Figure 6 three years after initial alignment.

(0.005 mm). This is a satisfactory tolerance. Over a period of eight years, 448 feet (137 meters) of rail stock were purchased. Fifty-three measurements of the diameter of these rails, ranging in length from 3 to 23 ft. (0.91–7 m) gave an average diameter of 2.4987 in. (63.467 mm) with a standard deviation of 0.0003 in. (0.008 mm). Thus, it is concluded that the rails, as such, have diameter tolerances within the alignment tolerances defined in Figure 4.

4.2.3. Riders

In 1972, it was decided to produce a quantity of riders at once to provide the standardization necessary for component interchangeability. Prior to that time an aluminum tooling plate had been used to fabricate riders on an "as-needed" basis, itself a lengthy process. Although mass production of the riders would be a typical process in an industrial production shop, it posed an unusual situation here, restricting the size of the first lot to 50 units. Using preaged, cast-iron forms, a procedure was developed to mill and surface grind the castings into a family of toleranced riders. Forty-five satisfactory riders were fabricated in approximately six weeks by four shop personnel working in collaboration with six graduate students. As a final fabrication step, the V and flat contact surfaces on each rider were hand scraped.

The riders were measured by placing them on standard rounds, approximately 1 ft. (305 mm) long, on a flat, granite surface plate. A depth micrometer was used to measure the height of the rider surface just to the inside of each corner. The dimension recorded was compared to the expected 1.000 in. (25.40 mm); if it fell within 0.010 in. (0.25 mm) of this, then the rider was considered acceptable. These measurements (for 45 riders) are shown in inset A of Figure 8 and summarized in inset B. A grand average of 0.9940 in. (25.248 mm) and standard deviation of 0.0028 in. (0.071 mm) was obtained for the 180 measurements. Inset C shows three contour lines for a single rider, indicating the relative flatness resulting from surface grinding.

On the basis of the experience gained in surface grinding, and the above data, it is expected that any local surface irregularity in a single rider will be within the range of the corner heights as indicated by the histograms in Figure 8. To date, this has been verified when devices were mounted on these riders.

4.2.4. Additional Beds

The remaining consideration in transfer of the primary ERL to other experiment beds throughout the laboratory is to have several beds available and aligned. To date, there have been two additional designs used for support pedestals. The main criteria involved in these designs were to achieve maximum rigidity over long lengths, to have sufficient bulk to the pedestal to allow secondary ERLs to be set at right angles to the primary ERL, and to build the pedestals such that they did not hinder the alignment of both primary and secondary ERLs. In addition, the open structure shown in Figure 4 was kept when practical to allow room below the rails and ERL.

A section of one pedestal is shown in Figure 9. This is an open truss-type structure. Four aluminum angles (e.g., part 1) are bolted together to form a four-sided pyramid, with the apex of each pyramid tucked into a 60° V angle trough (part 2) and the base into another set of V-shaped side supports (part 3). The pyramids are fabricated as individual, modular units, approximately 3 ft (0.9 m) on a side, and then bolted together with

A

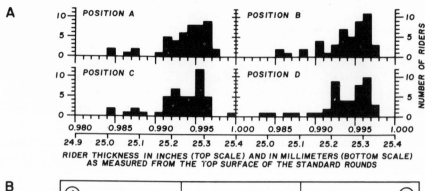

RIDER THICKNESS IN INCHES (TOP SCALE) AND IN MILLIMETERS (BOTTOM SCALE)
AS MEASURED FROM THE TOP SURFACE OF THE STANDARD ROUNDS

B

(A) 0.9940 ± 0.0030 in 25.247 ± 0.076 mm		0.9939 ± 0.0029 in 25.245 ± 0.074 mm (B)
	(FOR ALL POSITIONS A-D) 0.9940 ± 0.0028 in 25.247 ± 0.071 mm	
0.9934 ± 0.0029 in 25.232 ± 0.074 mm (C)		0.9940 ± 0.0032 in 25.248 ± 0.081 mm (D)

SUMMARY MEASUREMENTS FOR THE FOUR POSITIONS SHOWN
AVERAGE ± STANDARD DEVIATION FOR
RIDERS WITHIN 0.980 - 1.000 in RANGE

C

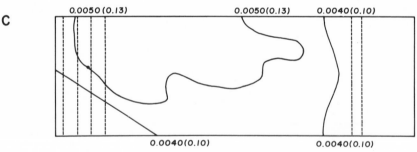

THREE CONSTANT-THICKNESS CONTOUR LINES

D

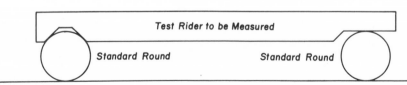

Granite Surface Plate

Figure 8. Dimensional tolerances obtained in the fabrication of 45 cast-iron riders. Using a test jig as shown in D, the rider thickness (Figure 3, dimension D) was measured at the four locations A–D as circled in B to produce the data shown in histogram form in A, or as averages in B, or as contour lines in C.

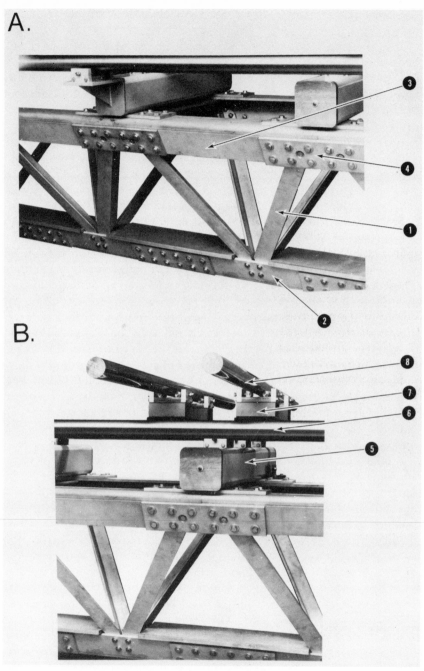

Figure 9. The basic, inverted-pyramid module used to assemble a variable-length, open-truss support pedestal for a long, rigid experiment bed.

flat "butt plates" (part 4) to form a rigid truss structure of the desired length.

Steel channels (part 5) are placed on the top of the truss at regular intervals to hold the rail mounts and rails (parts 7 and 6, respectively). In the application shown in Figure 9, inset B, riders have been placed on the lower rails, rail mounts placed on the riders, and another set of rails (part 8) laid in the mounts to establish a secondary ERL at right angles to the primary one.

The first truss pedestal was made 24 ft (7.32 m) long by bolting eight pyramid modules together. To ensure maximum rigidity with variable loading, six self-leveling air pods were placed under the ends of six of the steel channels to support the truss. Effective vibration isolation was also achieved in this manner.[17] Loading-deflection measurements on the truss pedestal, made while it was carrying a full set of rails, produced an elastic deformation of 0.0005 in. (0.013 mm) per 100-lb central load when the full span was only supported at the extreme ends. When the truss is air-suspended at six points, the bending under nonpoint loads is insignificant. The long truss support proved successful as a pedestal, and an additional 10-ft (3-m) unit was constructed.

Another type of support pedestal constructed involved the conversion of a large, steel-beam optical bench to a trirail bed. The beam was originally designed by Prof. Louis J. Gosting (deceased) of the University of Wisconsin Enzyme Institute. It was fabricated by the Beloit Corporation, and is shown in Figure 10, inset A. The base (part 1) is approximately 30 ft (9.1 m) long. It rests on blocks (part 2) via three points, and the two blocks are connected with a long steel channel (part 3). The base is supported on concrete pads (part 6) that are independent of the building floor. A high-precision, trapezoidal bar (part 5) runs the length of the base.

Rails were attached to the steel base, as illustrated in inset B of Figure 10, using an "outrigger" design. Aluminum I beams (part 2) were used to carry V blocks to hold the three sets of rails (part 3). The manner in which the six rails were aligned is shown in Figure 11. The I beam (part 1) is fastened to a plate (part 2) that fits up against a machined edge (part 3) on the main beam. At the other end of the beam, a rotating "U" joint (part 4) allows the beam to pivot according to the length of the support arm (part 5). The lower end of the support arm has a cone-and-socket connection (part 6) to the main beam. The length of the support arm is adjusted by rotating the threaded cone end (part 7). I beams are placed on either side of the trapezoidal bar (part 8). Once adjusted, they are locked together with a holding plate (part 9). The adjustment is straightforward, although time-consuming. All V blocks are machined to have common dimensions, simplifying the task of putting the rails on coplanar centerlines.

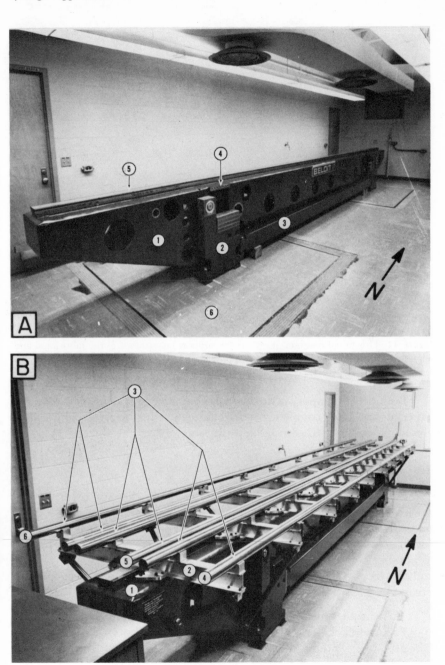

Figure 10. Fabrication of an aligned and fully load-bearing "tri-rail" experiment bed using a large steel beam as the central part of the support pedestal.

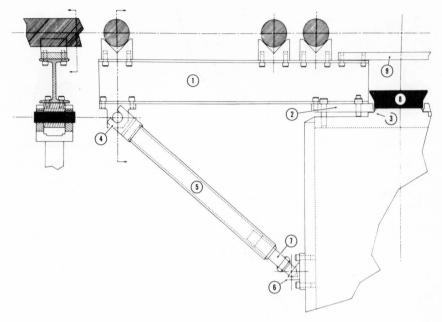

Figure 11. Functional diagram of one of the support beams used in the fabrication and alignment of the trirail experiment bed shown in Figure 10.

Both of the support pedestals described meet the required continuity expectations. They complete a last requirement for transferring a single, common primary ERL throughout the laboratory.

4.3. Working Tolerances

Based upon tolerances achieved for common dimensions of the rail stock and cast-iron riders, plus the successful alignment of other rail systems on pedestal bases similar to those shown in Figures 9 and 10, it is practical for components used for research on one experiment bed to be transferred to another bed, or another location on the same bed. They will coincide with the standard primary ERL to within the following typical accumulative error:

1. A rail-alignment error of ±0.001 in. (±0.025 mm)
2. A rail-diameter error of ±0.0005 in. (±0.013 mm)
3. A rider-flatness error of ±0.002 in. (±0.05 mm)

In addition to these errors, there will be those due to machine error in fabricating mounting mechanisms to hold components on a rider and others due to human enthusiasm. It is expected that with routine milling-machine tolerances of ±0.002 in. (±0.05 mm), and adequate time provided to use the transfer bed shown in Figure 6, all experiment beds in the laboratory will continue to allow component assembly or interchange to any primary ERL with a total error in the range of ±0.006 in. (±0.15 mm).

5. Components for Instrumental Systems

A wide variety of components may be assembled on an experiment bed. Once a component has been fabricated and set to the ERL, it may be placed on, moved along, or removed from a bed as the work being done on a bed dictates. To illustrate this, a few components traditionally associated with optical spectroscopy are presented here. The presentation should not be viewed as restrictive, but rather as illustrative. Available space precludes a full description of all components presently at hand. Neither should the specific support pedestal accompanying the component illustration be interpreted as its preferred location. Rather, the particular bed or pedestal shown was carrying the component at the time the photograph was taken. After discussion of these selected components, some of their possible arrangements into a system of instruments will be shown.

5.1. Spark Sources

In describing these components, the format of an instrument used for research to study the light emitted from a spark discharge has been followed. Thus, in the traditional approach, the first part of the instrument would be a "spark stand." Often a spark stand, because of its size, must be made an external part of a spectrometer, into which the electrodes are placed and from which light is emitted. However, because of the freedom offered by having room around the ERL and having the ERL locatable to the FMR independently of any "housings," etc., all light sources used here (including spark stands) may be made an integral part of a complete experiment.

5.1.1. Integral Source

For example, in Figure 12 is shown an example where the spark source, support frame (i.e., the stand), and the electrodes are integrated into a rider-based unit, through which light may pass. Rail mounts (part 1, inset B) hold rails (part 2) which in turn hold another rider (part 3) which holds more rail mounts (part 4) and another set of rails (part 5). This latter set of rails carries the primary FMR up into what would tradi-

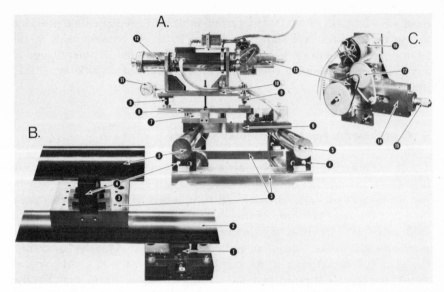

Figure 12. Example of a complete spark source constructed such that the source, "stand," and computer-driven (rotation) electrodes are integrated into a single, rider-based component.

tionally be the stand (the FMR is not lost). Then, in inset A, another rider (part 6) carries a rotating bearing (part 7), which supports a plate (part 8) onto which a dovetail slide (part 10) is resting via a three-point kinematic system of rounded screws (part 9). The resonant cavity of the quarter-wave[18] spark source (part 12) slides at 90 degrees to the ERL as measured by a dial indicator (part 11). All of these functions combine to allow precision drive of a sample electrode (assembly part 13, inset C), in translation to the ERL by a dovetail slide (part 14) with micrometer adjustment (part 15) and in rotation by a computer-driven stepper motor (part 16) and precision timing-belt link (e.g., part 17 and equivalent gearing).

The operating principles of this spark source have been presented in a prior publication.[19] However, it is germane that the spark and rotating electrode are positioned onto the ERL, not the converse. In this application, this was a secondary ERL, displaced upward from the primary ERL to allow other optics to direct light through the spark gap for absorption (thus the secondary FMR rails, part 5). By the simple expedient of picking up rider 6 and moving it from rail 5 to rail 2, the spark and rotating electrode are precisely located on the primary ERL, without loss of referenced optical alignment. In other words, it is the electrodes that are integral to the source, and the entire source (but for a radio-frequency power supply) that is integral to the experiment by virtue of primary and secondary ERLs and FMRs.

5.1.2. Experimental Source

In some work, integration of electrodes into the experiment via a specific source is premature because the electrodes themselves are still under development. In this case, an experimental "electrode support system" is used, which allows internal room for, and electrical connection with, any of a variety of prototype electrodes and associated mechanical and electrical hardware. One of three such systems is shown in Figure 13. It illustrates well the research value of the open ERL and rigid FMR resulting from the two-rail experiment bed.

Here (inset A) the primary rails of the bed (parts 1 and 2) carry first a short set of "cross-rail" (part 3) which are notched with a V and flat to act themselves as a rider (see also insets B and C, parts 1, 2, 3, and 4, and cross-tie parts 15 and 16). The cross-rails carry still another short set of notched rails (part 4 in insets B and C). The base of the electrode support system (part 5) is notched with a rotatable V and fixed flat. Thus, the whole base structure allows $X,Y,$ and θ motion about the ERL. In spite of

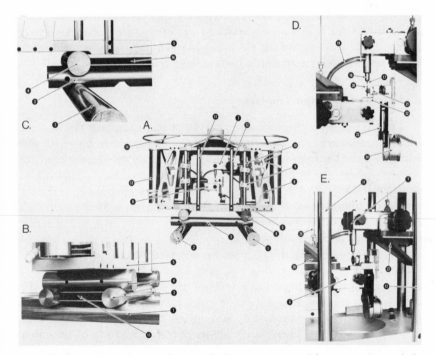

Figure 13. Assembly and detailed views of a large, computer-driven, experimental electrode support system used for research on electronic spark sources and the development of new electrode configurations.

the weight of the supported structure, these motions can all be made in small, precise steps, without cogging.

Experimental electrodes of considerable size and/or irregular shape are set and held on the ERL in two, stainless-steel, water-cooled jaws (parts 6 and 7). The jaws are precisely moved in the Z direction relative to the ERL by digitally driven stepper motors (part 8) coupled through ground ball screws to linear motion bearings (part 9) riding along precision-ground vertical shafts (part 10). To aid in placing electronic components (e.g., energy-storage capacitors) in the immediate vicinity of the spark, pairs of vertical rounds (parts 11 and 12) are placed in front and in back of the suspended electrodes. To allow study of heavy electrodes, the jaws are rigidly supported by gussets (inset E, parts 26 and 27) which are hooked by flexible steel wire (part 25) to adjustable counterweights hidden in circular housing at the extremes of the structure (inset A, part 13).

One experimental electrode configuration now under study (inset D) involves delivery of chemically reactive gases to the discharge through a flow nozzle (parts 17, 18, and 19). The sample electrode is a metallic disk (part 20) rotated on a bearing-supported shaft (parts 21 and 22) by a belt-coupled (part 23) stepper motor (part 24). The entire configuration may be moved about the ERL. This electrode support system is sufficiently open and versatile that it may contain, in addition to the electrodes, image-transfer optics and most of the electronics and circuit components required to form and deliver current to a high-voltage spark.

5.2. Mirrors for Image Transfer

Transfer of light through an experiment often suggests the presence of lenses. However, the experiment beds are sufficiently large to allow folding of light through compound angles, contrary to conventional "optical bars" where the light path is typically constrained to a single straight line. Thus, the use of lenses (which suffer from chromatic and other geometrical aberrations) is avoided in favor of mirrors. Mirrors up to 12 in. (305 mm) diameter have been used here with precise adjustment relative to the primary and secondary ERLs.

For example, in Figure 14 is shown one type of mirror mounting for a 12-in. blank. The mirror (inset A, part 1) rests on four cams (parts 11 and 12) which are positioned in reference to a support rider (part 2). The rider carries a rotating flange (part 3) which is coupled to the actual mirror mount by a sine-bar link (part 4) and spring-loaded (part 5) differential screw (parts 6 and 8). This linkage allows precision rotation of the mirror in the plane of the EMR. The mirror is translated along the bed for focusing using a "push-rider" (inset B, part 13) and dial indicator (part 7).

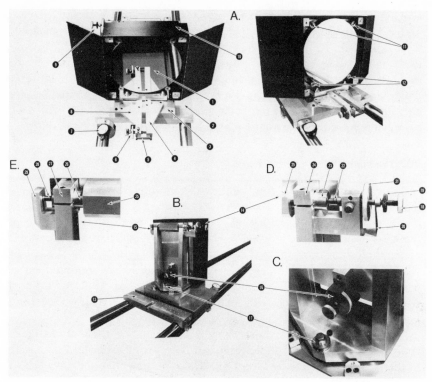

Figure 14. Mounting mechanism used to position and support variable-diameter mirrors on the ERL. Component parts shown in C, D, and E (see text) replace the traditional gimbal suspension.

In self-contained instruments where an open FMR is not available for ready measurement, mirrors generally are carried in gimbal mounts, such that orthogonal motions can be decoupled. Such gimbals are difficult to machine and largely redundant; once a mirror position is set it need not be continuously adjusted. Here, the gimbal mount has been dropped in favor of the sine-bar approach. With a constant FMR available, trigonometry may be used to bring about compensating adjustments, all measurable. By so doing, simplicity is achieved in the mirror mount design with high sensitivity in its angular adjustments. The size of those parts of the mount that could vignette the light path also is minimized.

For example, rocking the mirror (shown in Figure 14) is done from behind the mirror using a differential screw (part 16) to push against the bottom of a backing plate that holds the cams (parts 11 and 12) supporting the mirror blank. The same pivot bar that acts as a bearing for this vertical rocking motion (part 25, insets D and E) also slides in a V trough (parts 24 and 27) on its rounded ends (parts 23 and 26) to allow linear motion at 90° to the ERL. Another differential screw (parts 18 through 22) provides push against a spring-loaded plunger (parts 28 and 29) to provide this motion. In all, the mechanism is simple and sturdy. Mirrors of different diameters and/or focal lengths can be placed in the same mount by plugging other support cams into new positions on the backing plate. The completely mounted mirror may be moved between experiment beds.

5.3. Packaged Monochromators

A component that is not traditionally considered interchangeable, yet that still is only a part of a spectroscopic experiment, is a commercial packaged monochromator. While commercial monochromators are not transferred between experiment beds on a daily basis, they still are mobile. Here, three units have moved between four different beds in the last five years. The manner in which the monochromators are mounted on the beds to allow this mobility is indicated, by way of example, for a Jarrell-Ash 2.0-m Czerny-Turner unit in Figure 15 on the bed shown earlier in Figure 4.

This monochromator (inset A, part 1) is placed at the end of the bed, spanning two of the cast-aluminum support pedestals (part 2). By virtue of the pedestal support feet (part 3) and connecting channel (part 4) it was possible to place the monochromator on the prealigned bed without measurably altering either the relative heights and level of the machined top surfaces of the pedestals (part 5) or the linearity of the reference rail (part 6).

A sling mount is used to hold the monochromator (Figure 1) due to its particular manufacture. An aluminum plate (inset B, part 7) is slung beneath the rails by an inverted V and flat (V mount, part 8). A platform (part 9) is bolted to the bottom front plate (part 10) of the monochromator to contact the sling. This is accomplished by screwing a foot (part 12, inset C) into a block (part 11) to support the monochromator, and letting a swivel pad (part 15) on the bottom of the foot make the contact. This is done at three points, i.e., two in the front of the monochromator and one centrally located at the rear. By pushing on the bottom edge of these swivel pads with a screw (part 14), and adjusting the height of the foot by turning the threaded part (part 12), it is possible to position the monochromator entrance optical axis directly onto the ERL to within ±0.005 in. (±0.13 mm). In other words, the monochromator is aligned to the standard ERL, not the converse. A locking pad over the top of the foot (part 13) holds the monochromator in angular alignment to the ERL. Its entrance slit is positioned at the desired location along the ERL by sliding the sling mounts

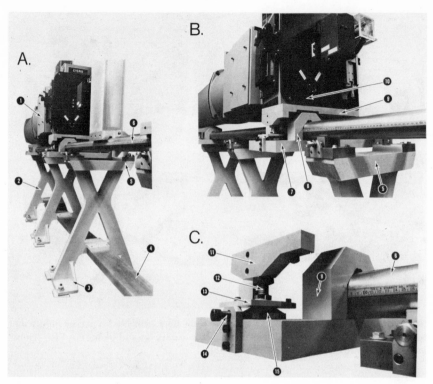

Figure 15. Use of a sling-mounted rider (Figure 1) to support and align a large, packaged monochromator on the ERL.

along the rails. A similar arrangement has been used to mount 1.0 m Jarrell-Ash and 1.0-m McPherson Czerny-Turner monochromators on other beds, whereas for small units (e.g., the McPherson model 700, 0.35-m Czerny-Turner) the monochromator is simply set on top of two conventional riders.

Use of a sling mount is a consequence of the open structure of the experiment beds, and probably could not be achieved with traditional optical bars and tables. The fact that a component as large as a monochromator can be moved along the ERL without losing alignment is a further advantage. The sling mounts used to hold the McPherson 1.0-m unit are shown in detail in Figure 16. Attention is directed to the dial indicator shown as part 8 of inset B. This indicator is clamped (part 7) to the reference rail, and marks the position of the monochromator along the ERL. When experiments require access to the focal plane normally occupied by the monochromator entrance slit, then the monochromator is slid back approximately 1 ft (0.3 m) and later repositioned. A repeatability as good as ±0.0005 in. (0.013 mm) has been measured for this repositioning.

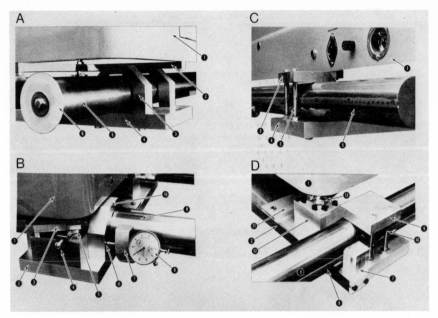

Figure 16. Assembly details of a pair of sling-mount riders suitable for precise, adjustable positioning of medium-sized, packaged monochromators on and along the ERL. Numbered parts are functionally identified in Table 2.

5.4. Specialized Components

5.4.1. Precision Slit

The wide separation of the two rails and the open structure between them offers further advantages for mounting other specialized-function components, as illustrated in Figure 17. In inset A is shown a slit assembly (part 1) whose tilt can be precisely set with a large spiroid gear (part 2) and whose position can be tuned with a precision micrometer (part 3). Because of the 15-in. (381-mm) separation between the rails holding the rider (part 5), there is room for both the large spiroid and sturdy gusset supports (part 4) to combine high-resolution angular adjustment with total component rigidity. Once adjusted, the entire slit assembly is quite mobile.

5.4.2. Cross Rails

A similar comment may be made about the single cross-rail shown in inset B (part 13). The height of this rail is determined by the thickness of the base rider (part 16), the thickness of the surface-ground steel "bricks" (part 15), and the adjustment made on the rail mounts (part 14). Because of the separation between the primary rails, the cross-rail may be up to 6 ft (1.8 m) long without tipping, allowing long cross-rails (e.g., Figure 7, inset B) to be used to generate secondary ERLs at various adjustable heights above the primary FMR. Such cross rails are fully load-bearing.

5.4.3. Auxiliary Light Source

In inset C is illustrated another type of auxiliary light source, the principles of which have been presented.[25] The entire source is housed in

Table 2. Detailed Parts of Figure 16

Item	Description
A: Rear Rider, Reference Rail Side	
1	1.0 M McPherson, Czerny-Turner monochromator
2	Bar to hold monochromator rear mounting foot
3	Inverted V brackets contacting reference rail
4	Sling-mount base running beneath the two rails
5	Fixed mechanical reference rail
6	Plate to stop rider motion at the end of the rail
B: Front Rider, Reference Rail Side	
1	Monochromator
2	Sling-mount base running beneath the two rails
3	C-clamp to lock monochromator foot to rider
4	Positioning screw pushing against foot edge
5	Height-adjusting, swivel-pad monochromator foot
6	Contact ball between dial indicator and rider
7	Movable rail-stop clamp holding dial indicator
8	Precision, 1.000-inch (25.40-mm) dial indicator
9	Fixed mechanical reference rail
10	Inverted V brackets contacting reference rail
C: Rear Rider, Auxiliary Rail Side	
1	Monochromator
2	Flat bracket contacting auxiliary rail
3	Bolts connecting flat bracket to rider base
4	Sling-mount base running beneath the two rails
5	Auxiliary rail
D: Front Rider, Auxiliary Rail Side	
1	Monochromator
2	Sling-mount base running beneath the two rails
6	Flat bracket contacting auxiliary rail
7	Optical vernier for coarse distance measurement
8	Machinist's tape for use with optical vernier
9	Auxiliary rail
10	Bolts connecting flat bracket to rider base
11	Flat bar connecting front and rear rider bases
12	Height-setting block to hold monochromator foot
13	C-clamp to lock monochromator foot to rider

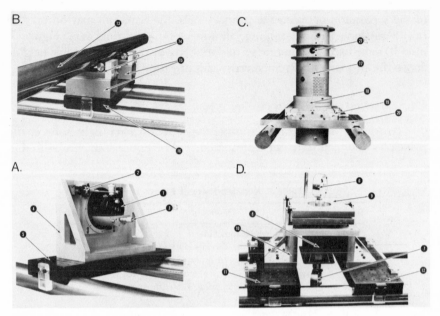

Figure 17. Samples of small, special-function, rider-based components that are used in various spectroscopic experiments.

a cylindrical can (part 17) that slides into an adjustable collar (part 18) gripped by a precision X–Y dovetail stage (part 19) in a rider (part 20). The rider has a hole bored through it, such that the electrodes contained in a demountable chamber at the top of the can (part 21) may be set on the ERL by sliding the can up and down inside the collar (part 18). The bottom of the can protrudes through the rider into the empty space between the rails. Many components use this principle to allow their size to be determined according to their function, instead of defaulted by available room.

5.4.4. Rotating Mirror

Another component using the empty space between and below the rails is shown in inset D. Here, an imaging mirror (part 6) is held in a collett which is attached to the rotor shaft of a hysteresis synchronous motor (part 8). The motor is fastened to a tiltable plate (part 9), which is supported by gusseted angle plates (part 10) between two riders (parts 11 and 12) whose top surfaces are coplanar. The other end of the motor shaft has a "trigger" mirror (part 7) fastened to it (see reference 10, Figure 5) to generate a pulse from a lamp and photocell mounted on the underside of the rider number 12 in precise synchronization to the position of the top mirror. The space between the rails and beneath the riders is valuable in allowing the top mirror to be positioned on the ERL independent of the motor length as such.

5.5. Mounted Diffraction Gratings

The use of two coplanar riders to support a single component (Figure 17D) illustrates how rigidity and interchangeability may be achieved when sufficient room is allowed in the early design phases of the experiment bed. Another example is shown in Figure 18, where a large diffraction grating (part 1, 102 × 256-mm ruled area) was placed on the ERL. Precision rotary motion was desired for this grating, but without the necessity of hanging high-tolerance, precision ball bearings. The "nest-of-balls" bearing shown in the inset was used.[20] The rotating ball was held in a shaft that was fastened to the grating yoke (one at either end), and the

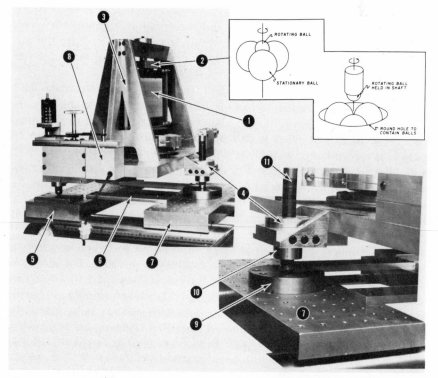

Figure 18. An A-frame mounting, using integral "nest-of-balls" bearings, for aligning and rotating a large diffraction grating relative to the ERL.

stationary balls were held in holes in plates above and below the grating. These plates were held together with A-frame gussets (part 3) to complete the bearing that supported the grating.

To allow the grating to be rocked and tilted onto the ERL, fine-pitched screws were used (part 4) to adjust the entire assembly. The screw (part 11) fit through a nut (part 10) into a socket (part 9) that rested on the surface of a rider (part 5 and 7). The two riders were connected with thin aluminum straps (part 6) to allow the assembly to smoothly slide along the rails. The grating was rotated without producing detectable image cogging of a test laser beam at the end of an optical level arm approaching 30 m by a digitally driven stepper motor (part 8). The mount shown was assembled and aligned to the primary ERL on the transfer bed shown in Figure 6 two years prior to its one-step transfer to the bed on which it is presently located. There was no loss of original alignment after the transfer.

5.6. Modular Vacuum Chambers

One experimental approach to exploratory research on the spark involves either extraction of species from the discharge into vacuum for mass sorting or study of its emission and absorption by vacuum ultraviolet spectroscopy. This requires that a relatively large volume around the primary ERL be evacuated, but without loss of the mechanical accessibility to the primary EMR.

To provide a vacuum about the ERL, modular vacuum chambers have been fabricated that are themselves mechanically referenced to the primary FMR, as shown in Figure 19. The chambers (inset A) are of two lengths, e.g., 1½ ft (0.45 m) and 3 ft (0.9 m). The small chamber in Figure 19 (part 1) is shown bolted to a large unit (part 2) by a large flange (part 7). Both chambers rest on riders. The riders set on a pair of notched cross-rails (part 4) that slide along two sets of coplanar primary rails (part 3). The two chambers may be bolted to a large flange on a liquid helium (or liquid nitrogen) cryopump (part 5) for evacuation without introduction of vibration. Small, liquid-nitrogen cryopumps are used for roughing.

The chambers are ported at three locations on each side and have a removable top plate to allow mechanical access to the ERL. For insertion of a spark source into the vacuum, a flexible bellows (part 6) is used to carry a large mounting flange (inset B, part 11) whose orientation relative to the ERL can be adjusted under vacuum with rotating bolts (part 12). Each chamber is kinematically mounted (part 10) to riders (part 9), allowing its orientation to be set with respect to the reference rail (part 8), i.e., the FMR.

This latter feature allows the chambers to be coupled at right angles to each other as well as end-to-end, i.e., to allow secondary ERLs to be

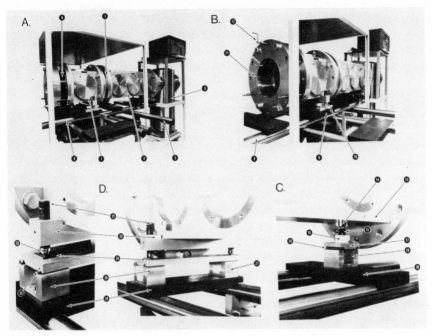

Figure 19. Modular vacuum chambers shown mounted with numerical reference to the ERL while connected to each other.

used. Each chamber is machined such that the surfaces of its end flanges are parallel and perpendicular to the surfaces of the side flanges. When mounted on the riders, the chambers then align to both each other and the rails. They vacuum seal without rocking off or twisting the aligned rails.

The mechanical alignment of each chamber is accomplished via its three mounting feet (insets C and D). A stud (part 14), fastened to the bottom of the chamber (part 13), holds a pivot ball (part 15) that sits in a cone socket (part 16) that rides on a small rider (part 17) and rails (part 18). The rails are on a small mount (part 19) resting on a conventional rider. This joint allows the rear of the chamber to swivel through an arc whose central tangent is at 90° to the reference rail. Inset D, parts 21 through 28, shows two equivalent swivel joints, except that one ball (part 23) rides in a V groove (part 24) instead of a cone, and the other simply rests on a hardened flat.

As is possible with monochromators, spark sources, diffraction gratings, and other large components, these vacuum chambers may be lifted off the rails on one experiment bed and transferred to another bed. With compact cryopumps available that may be bolted onto one of the side, end, or top ports of a chamber, it is possible to transfer "a piece of vacuum" to a portion of another experiment. To date, two large and one small chamber have been fabricated. Machine operations on a third large chamber are pending.

6. Experiment Systems

Since 1969, when the first experiment bed (Figure 4) was completed, several instrumental systems have been assembled from components similar to those shown here. Only two examples have been chosen to present the manner in which a system may be assembled from components. If it is recalled that all experiment beds have the same FMR and ERL, then this brief description introduces the experiment where the complete laboratory is called to function as one integrated instrumental system.

6.1. Time-Gated Spectrometry

The first complete bed-mounted system assembled here was based on the use of a rotating mirror (Figure 17, inset D) to sweep out a time-gated portion of the radiation emitted from a spark discharge. The assembly is shown in Figure 20.[21]

A short section (approximately 10 ft) of the truss-type pedestal (Figure 9) was used as the support base (inset A). The first step in the assembly was to place 13 riders on the rails to hold the components, including two sling-mounted riders (far left) to carry a McPherson 1.0-meter Czerny-Turner monochromator (Figure 16 shows these riders). The components were then added to the riders along two coplanar ERLs (insets B and C). One ERL ran directly down the center of the two rails (the primary ERL, labeled 19 in inset c). The other was displaced 5 in. (12.7 cm) to the side (the secondary coplanar ERL, labeled 20 in inset C).

The system functions in the followin manner. Light from or passing through the spark (held in part 14) is transferred via mirrors 12 and 13 through a three-mirror image rotator (part 11) to an intermediate, vertical "time-slit" (part 10). When the rotating mirror (part 9) is oriented as drawn, it optically connects the two ERLs (19 and 20). Light from the time-slit 10 is relayed via mirrors 8 and 9 to one of the two detectors. If folding mirror 6 is placed up, intercepting ERL 19, then light is folded into a photomultiplier (part 7) at 90° to the ERL. If mirror 6 is placed down, then light continues along the primary ERL to focus an image of the spark in the film plane of a 35-mm roll-film camera (part 5, insert B). If wavelength dispersion is desired, the camera and its rider are taken off the bed, and the monochromator (part 4) slid up along the rails until its vertical entrance

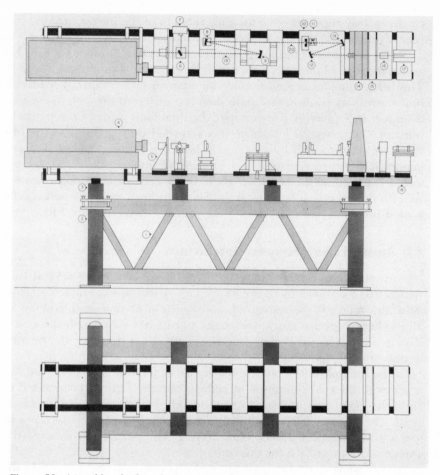

Figure 20. Assembly of selected components on a 3-meter section of open-truss bed to form an instrumental system for time-gated, spatially resolved emission spectrometry.

slit is in the focal plane (see Figure 16). The above happens only during the time that the two ERLs are connected as shown in inset C. Since mirror 9 is rotating at 3600 rpm the image of the vertical time-slit, 10, passes through the vertical entrance slit of the monochromator only for a brief period of time during each revolution of the mirror (if both slits are 0.050 mm wide, then the two ERLs are connected for about 0.1μs). A microcomputer is used to fire one spark in a train of sparks in precise phase synchronization to some particular angular orientation of mirror 9, such that monochromator 4 views and sorts radiation for a short time interval (i.e., a time gate) during the lifetime of the selected spark. The view is synchronously repeated once for each rotation of the mirror, making this system function as the spectroscopic equivalent of a boxcar integrator.

The presence of the image rotator along ERL 20 (inset C, part 11) causes mirrors 12 and 13 to appear to be orthogonal to mirrors 8 and 9. This brings about a cancellation of the astigmatism[22,23] introduced by all four mirrors. A point in the spark then is transferred through the system to appear as a point in, for example, the final focal plane of the monochromator. Thus, spatial information is recorded during the time the two ERLs are connected. If information is desired regarding the spatial distribution of absorbed radiation as a function of time, a pulsed "back-light" source is placed on the bed (part 16) to be viewed by the monochromator. As with other beds, a helium–neon alignment laser (part 17) periodically is used for numerical definition of the primary and secondary ERLs.

6.2. Atomic Vapor Absorption Spectrometry

Another system illustrating the transfer of light along several connected ERLs is shown in Figure 21.[17] This system was assembled to allow light absorption to be measured as a function of time and position in atmospheric-pressure spark discharges at a higher wavelength dispersion than available with the system shown in Figure 20. A dye laser may be used as one primary light source.

The multiple ERLs on this system are of interest. The main bed has a pedestal (inset A) composed of eight pyramids (Figure 9) connected to form the long truss previously discussed. The bed is suspended on six air pods. The primary rails run directly down the center of the truss (inset B) and are fully supported to be load carrying. They are used to establish the primary FMR and ERL for this bed.

Two, short secondary rails are laid coplanar with and parallel to the primary central reference and auxiliary rails. These provide secondary fixed mechanical references, but do not carry as much weight as the primary rails. This is in distinction to the trirail bed shown in Figure 10, inset B, where all three sets of coplanar rails are fully supported and load-bearing. The secondary rails are aligned to the primary FMR.

Following the procedure illustrated in Figure 9, inset B, and Figure 17, inset B, riders are positioned along the primary rails of this bed to support three sets of load-carrying cross-rails (inset C). Because these cross rails are carried on riders, they may be moved precisely and smoothly to locations anywhere along the primary rails. Thus, three tertiary, coplanar ERLs are set at right angles to the primary ERL, but may still be removed at any time.

When components are placed on the three ERLs, their utility becomes more evident (see inset D). Here, a spark (part 1) as held in, for example, the experimental support system shown in Figure 13, is placed on a tertiary ERL. Two mirrors (parts 2 and 3) relay light from the tertiary ERL to the primary ERL and through a Jarrell-Ash 1.0 m Czerny-Turner monochromator (part 4) which is sling mounted on the primary rails as shown in Figure 15. The monochromator functions primarily as a predisperser to the main 5.0 m spectrometer.

Light exiting the monochromator travels parallel to, but displaced to the side of, the primary ERL. It is recollimated by an 8-in. (203-mm) diameter, 5.0-m focal-length parabola, which is mounted on the primary rails in a manner similar to that shown for a 12-in. (305-mm) mirror in Figure 14. The angle of this mirror is adjusted to pass the light to the large, 590 groove/mm diffraction grating (part 6) shown previously in Figure 18. Typically, the grating is worked between the 10th and 15th orders. Dispersed light from the grating is subtended and focused by a camera mirror (part 7) in one of two focal planes.

If the folding mirror (part 8) is swung into the position shown in inset D, the dispersed light may be directed along a tertiary ERL to a rider containing one or more photomultipliers and exit slits (part 9), such that a "direct reader" results. Using rail stops (Figure 3, part 6) a family of such riders may be incorporated into the experiment for detailed study of different lines without changing the angular position of the grating. However, if the folding mirror (part 8) is swung in the opposite direction, light may be directed along the same tertiary ERL to, for example, a rider carrying a single, high-quality, adjustable slit (e.g., Figure 17, inset A). A high-dispersion monochromator then results as the grating is rotated.

To accomplish an absorption experiment,[10] present efforts are directed to the fabrication of a nitrogen-pumped dye laser. The nitrogen laser is to be placed on one of the secondary ERLs and the dye laser and cavity on a set of cross rails. Folding mirrors placed as shown would direct the output from the dye laser back through the spark, to retrace the same path as just described for light originating in the spark.

When operating the system shown in Figure 21, the laboratory room is darkened and all emitting light sources are masked to compensate for stray light. This avoids the need for restrictive "skins" and covers that inhibit monitoring the transfer of light through the experiment.

7. A Full Laboratory System for Synergic Action

Although only two of six experiment beds have been shown here carrying assembled systems, it is clear that all of the aligned experiment beds shown in Figures 4, 6, 9, and 10 can accept components and function as individual instrumental systems as soon as their primary ERLs are set.

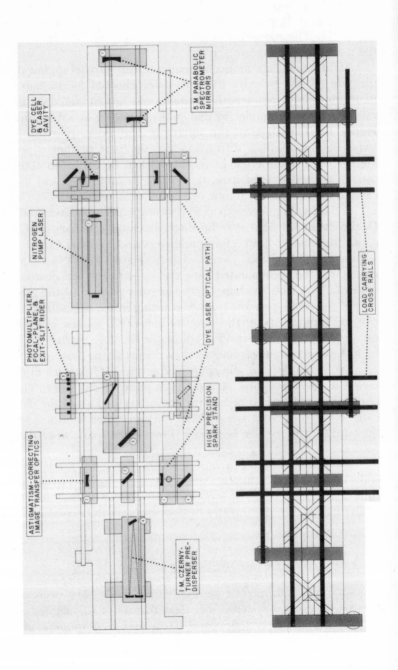

DYE CELL & LASER CAVITY

5 M PARABOLIC SPECTROMETER MIRRORS

NITROGEN PUMP LASER

PHOTOMULTIPLIER, FOCAL-PLANE, & EXIT-SLIT RIDER

DYE LASER OPTICAL PATH

ASTIGMATISM-CORRECTING IMAGE TRANSFER OPTICS

HIGH PRECISION SPARK STAND

LOAD CARRYING CROSS RAILS

I. M. CZERNY-TURNER PRE-DISPERSER

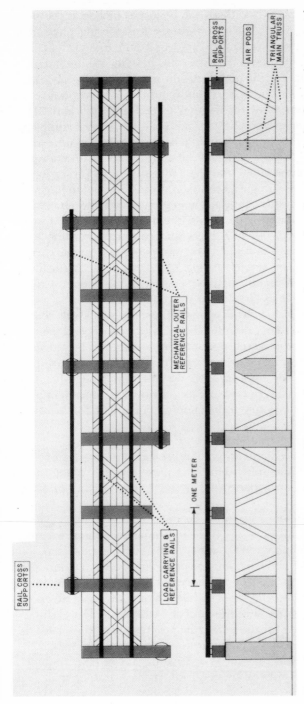

Figure 21. Establishing multiple ERLs and assembling selected components on an 8-meter section of open-truss bed to form an instrumental system for atomic vapor absorption spectrometry.

Since the components placed on them will be just as mobile as the height gauges and alignment laser shown in Figure 5 (considering physical size and weight only to be annoyances to rapid mobility), it is evident that a system assembled on any specific bed does not have to operate independently of the others. In fact, it may function as a subsystem of a larger instrumental system when the research indicates this is appropriate.

For example, when all of the systems (e.g., Figures 20 and 21) are constructed with the same type of fixed mechanical reference and aligned to the same common primary ERL, a light source producing information on one of them (e.g., Figure 12) may be transferred to another for a different specialized observation, and then returned to its earlier location for more work as part of a single experiment. It is equivalent to having the two instrumental subsystems connected together, except the observations are made on a time-sharing basis. Thus, the actual size of the research instrument is not limited, but may grow and contract according to the specific experimental needs at hand. (An analogy to the dynamic core size of a large computer that results during time-shared use according to demand would be appropriate.) In this perspective, the scope of the instrumental system may be as large as an entire laboratory.

The manner in which individual instrumental systems will combine into a laboratory-wide system cannot be described. It will depend too much upon the progress of the research problems under study. The individuals working on the problems will determine the number and types of instrumental systems that should appropriately operate simultaneously and interactively. By so doing, each person asserts individual, independent judgement and provides a necessary independent component for a synergic research experience.

Of major importance are the following:

1. The instrumental systems must allow unscheduled interaction. They must be able to function as subsystems on a time scale that is determined by individual research participants, not by their own fragility or limited functional lifetimes.

2. The interaction moments must be determined by an individual's perception of when it is appropriate. If the individual is learning, a learning curve will exist. Depending upon the input to this curve (its abscissa), some individuals may perceive interaction as appropriate sooner or more often than others. Some may never. The frequency or duration of an interaction cannot be predicted unless the input to individual learning curves is controlled, which is contrary to all of the overall goals in the approach.

3. The financial base for the individual systems must be as stable as is necessary for realistic interaction (neither token nor unrestrained).

4. The types of instrumental systems must be diverse enough to provide new, definitive physical/chemical information (rather than cyclic consistency).

5. The laboratory must have enough logistic organization to store past information for reasonably rapid and convenient recall.

Although an exact description of a synergic action cannot be provided here, the manner in which the above five points have been implemented to allow one or more can. The description that follows involves five experiment beds, arranged to facilitate exchange of information about the mechanisms of spark discharge. The description is aided with Figure 22, where the instrumental systems that would interact are designated in block form as subsystems of the larger, laboratory-wide, instrument system.

7.1. Interactive Subsystems

The experiments would be explored with, and fan out from, subsystem E, which has been shown in Figure 20. This assembly would provide quantitative information of where in the spark gap radiation was emitted, absorbed, or reemitted, covering a distance range of ±2.5 mm either side of the ERL with an accuracy of ±0.02 mm, absolute. It would also provide semiquantitative information as to when, and in what lines.

Information from E would be exchanged with subsystems A and D. Assembly A is set up on the bed shown in Figure 4 using the monochromator shown in Figure 15. It would provide quantitative information of how much light is emitted (or reemitted) within some specific location in

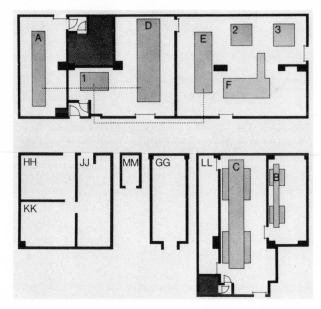

Figure 22. Integration of instrumental subsystems supporting facilities to produce a full instrumental system for interactive data exchange.

the spark. It also would provide semiquantitative information of how much light was absorbed by the spark. However, subsystem D, which was described previously in Figure 21, would provide quantitative information on how much light is absorbed within some specific location in the spark, also as a function of time.

To help bring about dynamic interactive data exchange between the above subsystems while they were simultaneously operating, they would be linked by a SERDEX (SERial Data EXchange) transmission system. They would be linked, first to each other and then to the University's Univac-type 1110 time-sharing computer. An available Princeton Electronic Products type 801 acoustically-coupled CRT graphics terminal (marked 1 in Figure 22) would act as a centrix for the video and data connection. This link would allow, among other things, data obtained at one subsystem to be telemetered to the 1110 for processing (e.g., Abel inversion[24] followed by Fourier transformation) and then returned for video and/or analog display at other subsystems. Via decoding electronics in each subsystem, the link also would allow the 1110 to set up control functions at each location, such as bursting stepper motors for grating rotation and the like.

Subsystem C is presently arranged as a single-mirror Schlieren camera (e.g., reference 26) and is assembled on the trirail bed shown in Figure 10. The data from this experiment, in the form of photographic maps of nonemitting species, would be used in designing experiments on A, D, and E. For this reason, C would be set up during the study for refractive index studies, although anomalous dispersion experiments could be conducted on the bed to combine the absorption data from D with refractive index changes in the spark.

Subsystem B is the transfer bed shown in Figure 6, and its functions have been described. As the SERDEX system would link A, D, and E electronically, so the transfer bed would link all subsystems optically and mechanically.

Subsystem F is presently still developing (the vacuum chambers shown in Figure 19 are being investigated on an "instrumental-research" basis). When mature, ion-extraction experiments would be executed on this bed to link the data from all other systems in terms of direct measurement of, among other things, the species responsible for current conduction in the central core of the spark.

The laboratory would function as a complete analytical instrumental system when the industrial-grade direct reading spectrometers, whose locations are indicated by blocks 2 and 3 in Figure 22, were complete.

7.2. Supporting Facilities

The supporting (i.e., service and back-up) facilities required to allow all of the subsystems to function as a complete system would be extensive and would require substantial evolutionary effort to develop. For example, the components not involved in a specific experiment would be cleared from the bed and stored in area LL. Here air quality and temperature would have to be controlled such that storage did not presuppose deterioration. Short wooden sections of rails fastened to the walls in this area would hold the rider-mounted components in a manner similar to that on a bed.

For almost all experiments, some form of photographic data registration would be used. To aid in uniform, standardized development and processing, two professional-caliber darkrooms (shown in dark shading in Figure 22) would serve all beds.

To allow mobility for light sources, detectors, and related electronic devices, it would be necessary to duplicate many of the digital control circuits at different beds. Because of the relatively high speed and low cost with which this may be done, it is a practical goal. To this end, a full-service digital electronics shop (area GG) would be maintained, including a small protected storage area (MM).

The logistical support problems for the system would be complex and intricate. For this reason, a full area (JJ) would be dedicated to purchasing, departmental and campus interactions, financial management, and publication. Laboratory records and data from previous experiments would be stored in area HH. Coordination would be provided (KK).

8. Progress

Much of the above description has materialized as presented. The arrangement of subsystems shown in Figure 22 is close to the present laboratory topography. Interactions of a diverse nature have occurred. The presence of continuity is evident on a component level, and the interchangeable instrument guideline has been achieved.

For example, the experimental electrode support system shown in Figure 13 has been associated with the research of two separate graduate students on two experiment beds over a period of eight years. It currently is a part of the interactive work of the professor on a third bed. The McPherson 1.0-m monochromator shown as part of the assembly in Figure 20 has been part of two different assemblies in the hands of three people. The Jarrel-Ash 1.0-m monochromator shown in Figure 21 has been part of two separate doctoral theses. The diffraction grating shown in Figure 18 was mounted and calibrated by one person using the transfer bed shown in Figure 6 and now is undergoing further modification, while it is being used by another on the assembly shown in Figure 21. The spark source shown in Figure 12 was designed and fabricated completely "by-the-numbers," i.e., from engineering drawings without being associated with, or fit to, any specific bed. Yet, it required only a small angular adjustment when its light output was recently recorded[19] using a system assembled on the bed shown in Figure 4.

From the viewpoint of experiment planning and retroactive design of expensive hardware, the interchangeability and lack of component obsolescence due to the common FMR and fixed-demension ERL have proven beneficial. For example, in Figure 23 is shown an early illustration of the first projected assembly of components for an ion-extraction experiment using the modular vacuum chambers shown in Figure 19. During the time that the first chambers were being constructed and assembled, parallel work on new spark sources produced a source[25] that was clearly better suited to this experiment than the quarter-wave unit indicated in the figure. Also, parallel work in schlieren studies of the spark[26] and first absorption experiments,[10] indicated that vacuum spectroscopy would be in order.

In light of these observations, work on the extraction experiment was put into what may be called a "holding pattern." Emphasis was shifted to accent spark-source development[27] and plans for fabrication of an additional chamber to be used eventually as part of the vacuum UV spectrometer were begun. In spite of these changes in direction, no hardware was scrapped, nor will it be. The possibility that the assembly indicated in Figure 23 may not materialize in the form shown is not a deterrent to the evolution of any of its components. They are, by virtue of their common dimensions, flexible.

8.1. Current Evaluations

From the viewpoint of permanence during quantitative measurement, of light intensities, the rigidity and long-term stability of components and beds have been successful. Experimental measurement tasks now can be accomplished in the daily laboratory environment that were not possible eight years ago.

For example, the rigidity and stability of the bed shown in Figure 21 makes practical open-camera, time-integrated photography of the light emitted from faint sources with spectral resolution of about 0.02 Å. Exposure times up to 16 h have been used. Also, the rigidity provided by the V-and-flat rail-rider combinations makes it practical to maintain the location of the accurately defined focal planes in the instrument system shown in Figure 20, for long periods of time between use or calibration. During use of this instrument, exposures up to 2 h may be required to detect light with spatial resolution at the 0.02-mm level, with time resolution approaching 0.1 μs microsecond and spectral resolution at 0.4 Å. Such exposures also are practical.

From the viewpoint of allowing exploratory but quantitative measurements involving diverse components, first experiments have produced encouraging results. For example, it recently has been possible to quantitatively measure the intensity of light emitted from the spark source shown in Figure 12 using photoelectric (photon-counting) detection and the monochromator shown in Figure 15. Details are reported in reference 19.

Some interactive experiments have occurred between instrument systems in a manner similar to that suggested by Figure 22. Initial schlieren data from a spark study prompted high-resolution absorption explorations, which were accompanied by spatial explorations in emitted and absorbed light. The net sum observations motivated further schlieren study, but at higher spark-repetition rates. The combined series of experiments strongly suggested that the time interval between the discharges in a train of sparks was an important experimental variable in determining the type of light emitted from any one spark in the train. This was not explored in work done before the first set of schlieren experiments.

Continuity also was demonstrated effectively in the above series of experiments. Prior spectroscopic[28] observations made in 1969 assisted in the design of those made in 1976. Schlieren photographs (and, indeed, the entire schlieren instrumental system) obtained between 1973 and 1974 were used to assist in obtaining the two sets of schlieren photographs in 1977. Three separate sets of laboratory records from past graduate students were used in the design of the optical/mechanical parts of the experiment. Four graduate students shared accumulated and current expertise on computers, digital electronics, and spark sources needed to execute the laboratory parts of data acquisition.

8.2 Directions

Perhaps the feature of the instrumentation illustrated here that will do most to achieve the overall goals through synergic action will prove to be its stability. Since the outcome of synergic research cannot be predicted by isolated study of individual participants, it follows that it cannot be forced, scheduled, or made accountable in advance of the fact. If the outcome is to be consequential, then the instrumentation facilitating the interactions must be able to weather the extreme conditions of no use to crowded use, sloppy experiments to refined measurements, isolated studies to collaborated project. In all cases, the foundations of the instrumentation must remain flexible for reworked experiments, by virtue of their stability. Such would appear to be the case.

While the experiment beds and instrument systems reported here are relatively young, they have received some unintentional, yet revealing,

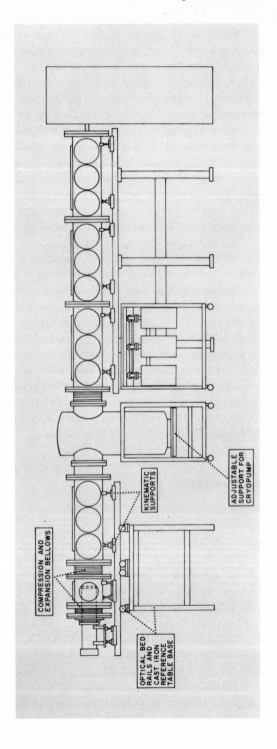

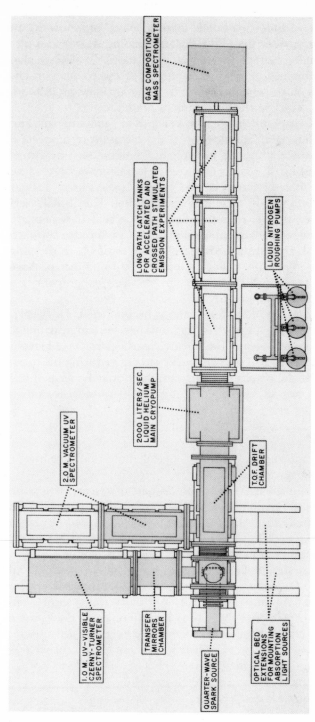

GAS COMPOSITION MASS SPECTROMETER

LONG PATH CATCH TANKS FOR ACCELERATED AND CROSSED PATH STIMULATED EMISSION EXPERIMENTS

LIQUID NITROGEN ROUGHING PUMPS

2.0 M. VACUUM UV SPECTROMETER

2000 LITERS/SEC. LIQUID HELIUM MAIN CRYOPUMP

T.O.F. DRIFT CHAMBER

1.0 M. UV-VISIBLE CZERNY-TURNER SPECTROMETER

TRANSFER MIRRORS CHAMBER

QUARTER-WAVE SPARK SOURCE

OPTICAL BED EXTENSIONS FOR MOUNTING ABSORPTION LIGHT SOURCES

Figure 23. An early projected layout of modular vacuum chambers (Figure 19) to allow spark ion extraction, observation, and manipulation.

stability tests. These include two floods (one of major proportions), two temperature cyclings greater than 20°F (loss of building steam), and up to three years of untended storage (student matriculation). In all tests, alignment either was maintained or restored in a straightforward manner. Only minor modifications have been necessary to bring up new projects with new participants by recycling the apparatus of predecessors.

In the study of spark discharge, the key link in catalyzing full interaction between contemporary participants is the inclusion of modern, direct-reading spectrometers into the instrumental system. Such instrumentation allows local testing of various mixtures of fundamental observations in analytical terms that directly relate to existing procedures. Tangible consequentiality indices rapidly arise from such testing. They usually serve to motivate whatever degree of interactive (new) research is perceived necessary to bring the total analytical situation under predictable and direct experimental control. One such instrument is now available to the effort and is being integrated into the rail/bed approach. As with other packaged devices, the spectrometer will retain mobility between experiment beds, without loss of its traditional integrity.

Synergic directions, by definition, cannot be predicted. However, if during the next decade the approach communicated here in instrumental terms provides an elected means for research participants to achieve inventive discovery in the course of their work, without reducing the quality or analytical utility of the results, then the approach may be termed technically effective. Its overall effectiveness must await outside assessment.

ACKNOWLEDGMENTS

It is difficult to appropriately acknowledge all those who have invested money, time, and energy in this effort.

The financial base for the work arose as part of research supported by the National Science Foundation under grants GP-7795, GP-13975, GP-35602X, and CHE76-17557. Additional funds were provided by the Research Committee of the University of Wisconsin Graduate School, the Department of Chemistry, and the Alfred P. Sloan Foundation.

Special attention should be directed to the graduate students and professional people who worked on the design, fabrication, alignment, and calibration of most of the instrument beds and components shown here. When the students contributed their ideas and time, it was done in conjunction with other educational requirements and research activities that often were not related to this work. Their investment is particularly noteworthy in that much of it was made when it was unclear if the concept or hardware would be more than an academic exercise.

In terms of their primary instrumental and/or laboratory contributions, these people are, in alphabetical order:

John S. Beaty (Figure 15)
Patricia Brinkman (Figure 22, areas JJ and KK)
David M. Coleman (Figures 9, 17, 20, 21, and 22, area GG)
William S. Eaton (Figures 16 and 20)
Steven A. Goldstein (Figure 20)
Bruce D. Hollar (Figures 6 and 18)
Jimmy W. Hosch (Figures 4, 6, 10, 14, and 15)
Robert J. Klueppel (Figures 20 and 22, darkrooms)
Jordan Lampert (Figure 22, areas B and LL)
John R. Rentner (Figures 12 and 22, area GG and darkrooms)
Alex Scheeline (Figure 22, item 1 and area HH)
Teruo Uchida (Figure 12)
Robert L. Watters, Jr. (Figures 19 and 23)

James B. Peters, mechanical engineer, worked on the design of much of the apparatus throughout the period 1968–1977, and in particular aided in the design of the first experiment bed (Figures 3 and 4). Substantial contributions also were made on the apparatus in Figures 11 and 13.

Particular appreciation is expressed to Russell Riley, Robert Schmelzer, Robert Lang, Kenneth Spielman, William Meier, Ronald Koopman, Thomas Lloyd, and Leo Rogers (deceased), all of the Chemistry Department Instrument Shops, for their enthusiastic support of this approach and high-quality efforts in the fabrication of a seemingly endless number of specific components and instrument beds.

Appreciation is expressed to Professors Paul Bender and J. W. Williams, Chemistry Department, University of Wisconsin, for making available to this effort the concrete pedestal shown in inset C of Figure 6. It was originally cast by Professor Williams in 1963 to support an ultracentrifuge. The cooperation of Professor James W. Taylor in allowing vacuum ultraviolet exploration studies on one of his research spectrometers is acknowledged.

Particular acknowledgment is directed to the Fisher Scientific Company through its Jarrell-Ash Division for awarding to the effort a Jarrell-Ash, model 750 Atomcomp Direct-Reading Optical Spectrometer. The continued encouragement for this approach during its instrumental development by Mr. John Bernier and Mr. Fred Brech, Jarrell-Ash Division of Fisher Scientific, Waltham, Massachusetts, and Mr. John Norris, National Bureau of Standards, is noteworthy.

Appreciation is expressed to Mrs. Pat Brinkman, Prof. David Coleman, Chemistry Department, Wayne State University, Detroit, Michigan, and

Prof. Merle Evenson, Department of Medicine, University of Wisconsin, for their valued assistance in the publication of this work.

Final acknowledgment is directed to my wife, Barbara, and parents, Lester and Phyllis, for their confident, patient, and understanding endorsement.

9. References

1. David B. Hertz and Albert H. Rubenstein, *Team Research*, p. 1–36, Eastern Technical Publications, New York (1953).
2. Leonard A. Herzenberg, Richard G. Sweet, and Leonore A. Herzenberg, Fluorescence-activated cell sorting, *Sci. Am., 234,* 108–117 (1976).
3. T. A. Mutch, S. U. Grenander, K. L. Jones, W. Patterson, R. E. Arvidson, E. A. Guinness, P. Arvin, C. E. Carlston, A. B. Binder, C. Sagan, E. W. Dunham, P. L. Fox, D. C. Pieri, F. O. Huck, C. W. Rowland, G. R. Taylor, S. D. Wall, R. Kahn, E. C. Levinthal, S. Liebes, Jr., R. B. Tucker, E. C. Morris, J. B. Pollack, R. S. Saunders, and M. R. Wolf, The surface of Mars: The view from the Viking 2 lander, *Science, 194,* 1277–1283 (1976).
4. H. V. Malmstadt and C. G. Enke, Electronics and instrumentation in chemical research, *Anal. Chem., 33,* 23A–27A (1961).
5. Jeanette G. Grasselli, Teaching analytical chemistry: Real world needs, *Anal. Chem., 49,* 183A–192A (1977).
6. Benjamin S. Bloom (ed.), *Taxonomy of Educational Objectives,* p. 162–184, David McKay Co., New York (1956).
7. J. P. Walters, Historical advances in spark emission spectroscopy, *Appl. Spectrosc., 23,* 317–331 (1969).
8. J. P. Walters and H. V. Malmstadt, Emission characteristics and sensitivity in a high-voltage spark discharge, *Anal. Chem., 37,* 1484–1502 (1965).
9. J. P. Walters, Short-time electrode processes and spectra in a high-voltage spark discharge, *Anal. Chem., 40,* 1540–1573 (1968).
10. David M. Coleman and J. P. Walters, Evidence for an ionic, toroidal, post-discharge environment in the atmospheric pressure spark discharge, *J. Appl. Phys., 48,* 3297 (1977).
11. J. P. Walters, Source parameters and excitation in a spark discharge, *Appl. Spectrosc., 26,* 17–40 (1972).
12. Richard R. Skemp, *The Psychology of Learning Mathematics,* p. 72–82, Penguin Books, Baltimore (1973).
13. John Strong, *Procedures in Experimental Physics,* p. 586–590, Prentice-Hall, New York (1938).
14. P. H. Abelson and A. L. Hammond, The electronics revolution, *Science, 195,* (4283), 1087–1091 (1977).
15. S. E. Harris, J. F. Young, A. H. Kung, D. M. Bloom, and G. C. Bjorklund, Generation of ultraviolet and vacuum ultraviolet radiation, in: *Laser Applications to Optics and Spectroscopy, Physics of Quantum Electronics* Stephen F. Jacobs, Murray Sargent, James F. Scott, and Marlan O. Scully, ed., Vol. 2, pp. 181–197 Addison-Wesley, Reading, Massachusetts (1975).
16. D. M. Coleman and J. P. Walters, Simple adjustable mounting for high-precision optical rails, *Appl. Spectrosc., 31,* 172–173 (1977).

17. David M. Coleman and John P. Walters, Design of a large, rigid, modular optical bed for versatile optical and spectroscopic experimentation, *Spectrochim. Acta, 33B,* 127 (1978).
18. J. P. Walters and T. V. Bruhns, Electronically-controlled, current-injection spark source for basic and applied emission spectrometry, *Anal. Chem., 41,* 1990–2005 (1969).
19. John R. Rentner, T. Uchida, and J. P. Walters, A 323 MHz quarter-wave spark source for the production of stable spark discharges, *Spectrochim. Acta, 32B,* 125 (1977).
20. J. P. Walters, B. D. Hollar, and D. M. Coleman, Simple bearing for high-precision grating rotation, *Anal. Chem., 48,* 1261–1263 (1976).
21. R. J. Klueppel, D. M. Coleman, W. S. Eaton, S. A. Goldstein, R. D. Sacks, and J. P. Walters, A spectrometer for time-gated, spatially-resolved study of repetitive electrical discharges, *Spectrochim. Acta, 33B,* 1 (1978).
22. S. A. Goldstein and J. P. Walters, A review of considerations for high-fidelity imaging of laboratory spectroscopic sources—Part I, *Spectrochim. Acta, 31B,* 201–220 (1976).
23. S. A. Goldstein and J. P. Walters, A review of considerations for high-fidelity imaging of laboratory spectroscopic sources—Part II, *Spectrochim. Acta, 31B,* 295–316 (1976).
24. Alexander Scheeline and J. P. Walters, Considerations for implementing spatially-resolved spectrometry using the Abel inversion, *Anal. Chem., 48,* 1519–1530 (1976).
25. D. M. Coleman, J. P. Walters, and Robert L. Watters, Jr., An integrated electronic spark source for basic and analytical emission spectroscopy, Spectrochim. Acta, 32B, 287 (1977).
26. J. W. Hosch and J. P. Walters, High spatial resolution Schlieren photography, *Appl. Opt., 16,* 473–482 (1977).
27. D. M. Coleman and J. P. Walters, An electronic, adjustable-waveform spark source for basic and applied emission spectrometry, *Spectrochim. Acta, 31B,* 547 (1976).
28. J. P. Walters and S. A. Goldstein, How and what to sample in the analytical gap, Special Technical Publication No. 540, Sampling, Standards and Homogeneity, pp. 45–71, American Society for Testing and Materials, 1916 Race Street, Philadelphia, Pennsylvania (1973).

4

Correlation Methods in Chemical Data Measurement

Gary Horlick and Gary M. Hieftje

1. Introduction to Correlation

Correlation techniques have long been used to measure and process signals in physics, chemistry, and engineering. Lee *et al.*[1] discussed the application of correlation analysis to the detection of periodic communication signals rather early and later reviewed the topic in some detail.[2] Correlation techniques, again applied primarily to communications, have also been discussed by Lange[3] and were applied at a relatively early stage to the analysis of electroencephelographic data.[4,5] Although correlation techniques have been employed in these and other fields with considerable advantage and success, their use has unfortunately been somewhat limited by a lack of understanding of the principles involved in the computation, utilization, and measurement of correlation functions. In addition, for practical purposes it has been difficult to use correlation techniques because effective methods and instrumentation were not available for the rapid, automatic evaluation of correlation functions.

In this chapter, we will attempt to present a treatment of correlation analysis that is both understandable and useful to the practicing scientist or chemist. Many treatments of correlation analysis lose the basic simplicity of the technique in an excess of mathematical equations and in applications to arbitrary waveforms. Throughout, a qualitative or semiquantitative view will be adopted, with a minimum amount of mathematical detail. While

Gary Horlick • Department of Chemistry, University of Alberta, Edmonton, Alberta, Canada T6G 2G2 **Gary M. Hieftje** • Department of Chemistry, Indiana University, Bloomington, Indiana 47401

some rigor will necessarily be sacrificed by this approach, we feel that the importance and utility of the ubiquitous correlation operation will be rendered most meaningful to the majority of readers.

Today, there is rapid growth in the utilization of correlation methods for the measurement of chemical data. Many computer signal-processing methods are based on correlation, and scientists are increasingly using techniques that are inherently based on correlation for the measurement and generation of chemical data. In large part, this upswing in the utilization of correlation techniques is due to developments in instrumentation.

A number of commercial hardware correlators are available, and the ready availability of digital computers in most laboratories makes software approaches to correlation both simple and relatively inexpensive. In addition, the rapid development of increasingly sophisticated electronic circuitry in small inexpensive packages can be expected to produce a number of new and unique approaches to the design of laboratory-oriented correlators. These advances in instrumentation and software are certain to generate further rapid growth in the utilization of correlation techniques in experimental science; consequently, it is essential that workers in all branches of science become increasingly aware of correlation operations, their use, and their application.

1.1. What Is Correlation?

Simply stated, correlation analysis provides information about the coherence within a signal or between two signals. The correlation function of two signals is obtained by evaluating the time-averaged or -integrated product of the two signals as a function of their relative displacement. Mathematically, correlation can be expressed as:

$$c_{ab}(\tau) = \lim_{T \to \infty} \frac{1}{2T} \int_{-T}^{+T} a(t)b(t \pm \tau) \, dt \qquad (1)$$

where $c_{ab}(\tau)$ is the correlation function between the two signals $a(t)$ and $b(t)$, and τ is their relative displacement. The signals can be a function of essentially any variable, e.g., wavelength, retardation, frequency, accelerating voltage, time, etc. Thus, if a and b are considered to be functions of time, the correlation function c_{ab} will be related to and plotted against the relative time delay between the two signals.

In most situations, correlation is implemented on digitized signals. For digitized signals, the calculation of the correlation function can be expressed by the following summation:

$$c_{ab}(n\Delta t) = \sum_{t} a(t)b(t \pm n\Delta t) \qquad n = 0, 1, 2, \ldots \qquad (2)$$

The signals can only be displaced some integral number of the sampling interval, Δt. Thus, the displacement $n\Delta t$ is equivalent to τ in equation (1).

Two different correlation operations can be identified. If $a(t)$ and $b(t)$ are identical (i.e., if $a = b$), an *autocorrelation* function is obtained by application of equation (1). Thus, autocorrelation indicates whether coherence exists within a signal. In contrast, a *cross-correlation* function, produced if $a(t)$ and $b(t)$ are different, shows the similarities between the two signals.

1.1.1. Autocorrelation

To illustrate the process of correlation, let us consider a simple instrument capable of producing a correlation function. From equation (1), correlating two signals that are functions of time merely involves multiplying one of them [$a(t)$] by a delayed version of the other [$b(t - \tau)$], averaging the product, and expressing the averaged value as a function of the chosen delay. An instrument capable of performing these procedures is shown in Figure 1; its operation can be understood most simply by autocorrelating a simple sine wave.

If a sine wave is to be autocorrelated, it will have to be sent simultaneously into both inputs of the correlator. To begin, assume that $\tau = 0$; that is, that the sine wave is multiplied by itself *in phase*. When this occurs, the output from the multiplier will be a sine^2 wave, as illustrated in Figure 2a; this will produce an output from the averager equal to the mean square of the original sine wave. Thus as plotted in Figure 3, the value of the autocorrelation function at $\tau = 0$ will be the mean square of the original wave.

Now imagine that the variable delay in the autocorrelation computer (Figure 1) has been increased (manually or automatically) by an amount of time equal to one-quarter period of the original sine wave. Under these conditions, the sine wave will be multiplied by a version of itself which is

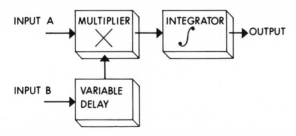

Figure 1. Block diagram of a simple correlator.

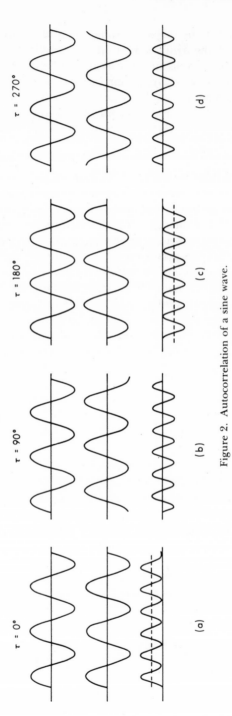

Figure 2. Autocorrelation of a sine wave.

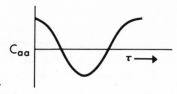

Figure 3. Autocorrelation function of a sine wave.

effectively phase-shifted by 90°. As shown in Figure 2b, this multiplication will generate a wave having an average value of zero. This value has also been plotted in Figure 3.

The variable delay is now increased to a time equal to one-half period of the input sine wave to produce a situation depicted in Figure 2c. The averaged multiplier output is now just the negative mean square of the sine wave and is again shown plotted in Figure 3. Increasing the delay another quarter period causes the original wave to be 270° out of phase with its delayed replica, so that, as shown in Figure 2d, the product again has a zero average.

Finally, when the variable delay has been increased to a time exactly equal to one period of the input sine wave, the wave and its delayed replica will once more appear to be in phase, to produce a situation identical to that when the delay (τ) was zero. From this and Figure 3, it should be apparent that the autocorrelation function of a sine wave is itself sinusoidal but has its maximum value shifted to coincide with $\tau = 0$. Mathematically, this means that the sine wave, of indeterminate phase, has been converted to a cosine wave. Thus, if the original wave were expressed as

$$a(t) = A \cos(2\pi f t + \theta) \tag{3}$$

where A is the maximum amplitude of the sine wave, f is its frequency, and θ its phase, the autocorrelation function of the wave would follow the relation:

$$c_{aa}(\tau) = (A^2/2) \cos(2\pi f t) \tag{4}$$

From this result, Figure 3, and the preceding discussion, several important characteristics of the autocorrelation function should be evident. First, the autocorrelation function is even, that is, it is symmetrical about the point $\tau = 0$. In effect, this is identical to saying that either of the two channels of the autocorrelator of Figure 1 can contain the variable delay network, and that changing the delay to the opposite channel has the mathematically interpreted consequence of using a negative delay. Secondly, the amplitude and frequency (or period) of the autocorrelation function are unambiguously related to those of the original sine wave. The period of the two are, of course, identical while the amplitude of the autocorrelation function is just the mean square of the original. A third property of the autocorrelation function of Figure 3 is its infinite length.

Any periodic function will display a similar characteristic since any such waveform is just a combination of a number of constituent sine waves, each with its own amplitude and phase characteristics. This periodicity or coherence, observed in most signals, is the feature which enables them to be separated so effectively from superimposed noise, which is ordinarily neither periodic nor coherent.

Finally, although frequency and amplitude information about the original waveform is carried in the autocorrelation function, the phase of the original signal is lost. This fact is emphasized by the autocorrelation function of a square wave, shown in Figure 4 and constructed just as for the sine wave of Figure 2. The triangle wave autocorrelation function shown in Figure 4 contains the same frequencies (i.e., sine wave components) as did the original square wave. However, in the triangular wave all the sine waves are in phase, to produce a different summation wave shape. In the understanding of the square wave autocorrelation, it is instructive to imagine the two square waves sliding past each other. The autocorrelation function is just the changing mutual area of the two functions.

A familiar example of a related autocorrelation is the scanning of a spectral line (image of the entrance slit) across the exit slit of a monochromator to produce the slit-width-limited triangular resolution function. This autocorrelation operation is illustrated in Figure 5.

Autocorrelation of nonperiodic or random (noise) waveforms produces markedly different results from those obtained for periodic waves. To understand this, we need only recognize that any wave, whether peri-

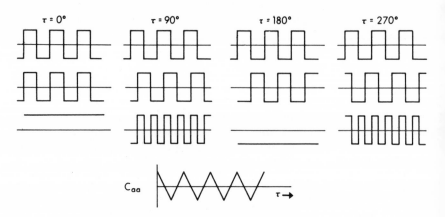

Figure 4. Autocorrelation of a square wave.

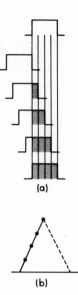

(a)

Figure 5. Autocorrelation of a rectangular pulse.   (b)

odic or not, is composed of sine wave components. Each of these components, when autocorrelated, will produce a cosine wave beginning at $\tau = 0$. When many such components are present in a parent waveform, the resulting autocorrelation will just be the sum of the autocorrelation functions of the components, as illustrated in Figure 6. A true random waveform (white noise) contains all frequencies. The cosine autocorrelation images of all these frequencies reinforce at $\tau = 0$ to produce a value equal to the mean square of the original random signal, and at any point beyond $\tau = 0$ they destructively interfere to produce a time-averaged value of zero (see Figure 7).

In order to obtain this ideal autocorrelation function of random noise,

Figure 6. Cosine autocorrelation images.

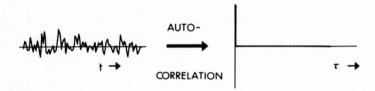

Figure 7. Autocorrelation of a random waveform.

the bandwidth of the noise must be infinite. Of course, no process in nature is truly random. Real random waveforms are "band-limited" and do not produce a single spike upon autocorrelation but instead a shape such as portrayed in Figure 8a. The functional form is a decaying exponential whose width is inversely proportional to the bandwidth of the noise waveform. Note that in Figure 8a, both sides of the autocorrelation function have been included to demonstrate its symmetry. In all other presentations, only one side of the autocorrelation function will be given; the reader should recall the actual symmetrical characteristic.

When Figure 8a is compared with Figures 3 and 4, it is clear why autocorrelation is such a powerful technique for the extraction of periodic signals from noise.[6,7] Most noise, being inherently random, contributes to the autocorrelation function only at very small values of τ. A periodic signal, in contrast, will continue to contribute even at very large values of τ. Therefore, to measure the amplitude and frequency of a periodic signal buried in noise, it is necessary to examine the autocorrelation function at a location well removed from the central ($\tau = 0$) peak. To clarify this point, the autocorrelogram (autocorrelation function) of a noisy sine wave is portrayed in Figure 8b.

The autocorrelation function also provides a very convenient means of determining the signal-to-noise ratio of a measured waveform. As illustrated in Figure 8b, the value of the autocorrelation function at $\tau = 0$ is just the mean square of the original signal plus noise, whereas the peak of the autocorrelogram at large values of τ is equal to the mean square of the

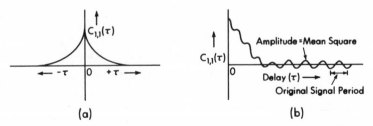

Figure 8. Autocorrelation function of band-limited noise (a) and of a noisy sine wave (b).

signal alone. Because the mean square value is directly proportional to the power of an alternating waveform, the signal-to-noise power ratio can be calculated from the mean square of the signal divided by the difference between the measured quantities (S + N and S).

1.1.2. Cross-Correlation

Cross-correlation analysis is similar in structure and concept to auto-correlation except that in cross-correlation the coherence between two waveforms is examined rather than the coherence within a waveform. From our discussion of autocorrelation, it should be clear that this coherence can only exist for frequency components which are common to the two waves. This is true because multiplication of two sinusoidal waveforms of different frequency produces a waveform which is itself symmetrical about zero and has a zero time average.

For example, if two waveforms to be cross-correlated are sinusoidal and of the same frequency, the situation is similar to that discussed earlier for autocorrelation. However, unless the two sine waves happen to be in phase, the cross-correlation function will not be symmetrical about $\tau = 0$, but will be displaced by an amount related to the phase difference. This is illustrated in Figure 9. For two waves of different shape but containing common frequency components, the situation is even more complex. A sine wave and square wave of the same fundamental frequency will thus cross-correlate to a sine wave of that frequency, which is the only component common to both. Furthermore, the shift of the cross-correlation function from $\tau = 0$ will once more indicate the phase relationship between that frequency component in the two waves.

With random or stochastic waveforms (i.e., noise), cross-correlation behaves quite differently from autocorrelation. In particular, because any two random waveforms are inherently independent, they will not correlate at any point, so that the cross-correlation function will everywhere be zero as the product of two random functions will itself be random and thus have a zero time average. This property is extremely important in signal-processing applications, where a noisy signal is to be detected with the aid of an available reference wave at the same frequency. Consider, for ex-

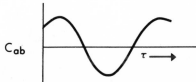

C_{ab}

Figure 9. Cross-correlation function of two sine waves.

ample, the detection of a noisy square wave signal using a noise-free sinusoidal reference wave. The only common feature between the signal and reference waves is that component at the reference wave frequency; this component alone will thus appear in the cross-correlation function. The cross-correlation function (at $\tau = 0$) will then have an amplitude equal to the averaged product of the noise-free signal and reference waves (see Figure 10).

It will be obvious to most readers that the cross-correlation operation depicted in Figure 10 forms the basis of the lock-in amplifier.[6-8] The only major additional component in a lock-in amplifier is a phase-shifting circuit, usually on the reference channel. This allows the output of the correlator ·to be maximized by ensuring that the signal and reference waveforms are in phase.

The fact that the amplitude of the cross-correlation function depends both on the reference and signal wave amplitudes can often be used to advantage in signal-processing applications.[9] Merely by increasing the magnitude of the reference wave, the cross-correlogram amplitude can be enlarged; this is clearly advantageous over the autocorrelation operation, where no reference wave exists and the final correlation function is only as large as the mean square of the original signal.

From the foregoing discussion, it should be apparent that correlation operations, while outwardly and mathematically a bit forbidding, are actually quite simple. Also, several important characteristics of both auto- and cross-correlation have been shown to be of importance to the detection of signals. The examples shown have primarily utilized periodic-type signals. A considerable number of signals of chemical interest consist of peaks. Several examples illustrating the application of correlation to the measurement and processing of peak-like signals will be discussed later. But first, two related areas must be covered, convolution and Fourier transformation. Some knowledge of these areas is important for a full understanding of correlation operations.

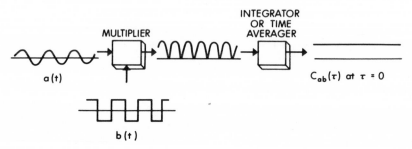

Figure 10. Cross-correlation of a sine and square wave (lock-in amplifier) at $\tau = 0$.

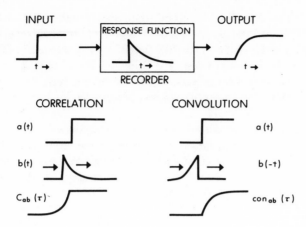

Figure 11. Convolution and correlation of a step waveform with an exponential instrument response function.

1.2. Convolution

A number of the aspects of correlation that have been introduced are perhaps more familiar to the reader under the term convolution. However, convolution, an operation which occurs during the generation or measurement of all signals, can be considered to be merely a special kind of correlation process. Mathematically, convolution can be expressed as

$$\text{con}_{ab}(\tau) = \lim_{T \to \infty} \frac{1}{2T} \int_{-T}^{+T} a(t)b(-t \pm \tau) \, dt \qquad (5)$$

If equations (5) and (1) are compared, their only difference is observed to be a minus sign in front of the t of the b function. The effect of this minus sign is to reverse the $b(t)$ function (from left to right) on the time axis before it is multiplied by $a(t)$. The rest of the shifting, multiplying, and averaging procedures are identical to those employed in correlation. Therefore, correlation and convolution are identical, except that in convolution, one of the signals is first reversed.

The reason for this can be best understood with the aid of an example. Consider a rapid signal change (step) being displayed on a recorder having a relatively long time constant. The recorded signal will, of course, not be identical with the original step but will reflect the response time of the recorder.[10] For simplicity, let us assume that the recorder's response is exponential, as shown in Figure 11. By convention both the input to the recorder *and its response function* are expressed with time increasing to the right. Therefore, if we imagine the input signal entering the recorder, it is apparent that the left side of the signal would enter first and encounter the response function. For this to occur properly, either the signal or the response function must be reversed. It is conventional to reverse the response function. Thus the recorded signal is then the *convolution* of the input step and the recorder's response function. Notice that, if a correlation between those two functions were produced by the recorder, the leading edge of the recorder's output would be curved upward rather than downward, as shown in Figure 11.

From these considerations, convolution can be seen to be merely a special kind of correlation and can often be implemented and utilized in similar ways. In fact, in those many instances in which the response function $b(t)$ is symmetrical, its inversion produces no change, so that correlation and convolution provide identical results.

1.3. Fourier Transforms in Correlation

As mentioned in the introduction, the ready availability of digital computers in most laboratories facilitates software approaches to correlation. In particular, the fast Fourier transform algorithm has essentially revolutionized the extent to which correlation techniques can be implemented on computers and indeed many correlation-based techniques are more commonly referred to as Fourier transform techniques. In addition to providing a convenient route for the implementation of correlation, a knowledge of Fourier transforms also aids the utilization and understanding of correlation operations. A very important theorem concerning correlation states that correlation of two waveforms is equivalent to multiplication of their Fourier transforms followed by inverse Fourier transformation of the product. Schematically, this sequence can be represented in the following way:

$$a(t) \quad * \quad b(t) \ = \ c_{ab}(\tau) \tag{6}$$

$$\downarrow \text{Fourier transformation} \quad \downarrow \quad \uparrow \text{Inverse Fourier transformation}$$

$$A(f) \quad \times \quad B(f) = C_{AB}(f) \tag{7}$$

The asterisk in equation (6) is merely a common shorthand way to denote correlation; equation (6) is therefore identical to equation (1). It should be noted that this shorthand form is used by many authors to designate convolution as well. However, in this chapter we will employ it exclusively to denote correlation.

Notice that in equations (6) and (7), cross-correlation is shown to involve nothing but a multiplication process in the Fourier domain. Simple as this aspect of correlation might seem, it is of paramount importance. A large fraction of software and hardware methods for obtaining the correlation function rely on the Fourier transform method because of its speed and relative simplicity. Let us consider the Fourier process in a bit more detail.

Fourier transformation is a mathematical means of relating two functions. It may be stated as

$$a(t) = \int_{-\infty}^{+\infty} A(f)e^{2\pi i t f}\, df \tag{8}$$

The inverse Fourier transform is

$$A(f) = \int_{-\infty}^{+\infty} a(t)e^{-2\pi i t f}\, dt \tag{9}$$

The functions $a(t)$ and $A(f)$ constitute a Fourier transform pair. Fourier transformation, as it is usually applied in chemistry, involves the determination of the frequency composition of a waveform. Often the waveform $[a(t)]$ will represent a time-varying property of a substance or device under investigation, in which case the frequency composition $[A(f)]$ will be expressed in units of s^{-1} or Hz. However, as discussed earlier with regard to correlation, a waveform being transformed can be a function of any variable.

The frequency composition of a waveform is ordinarily expressed in the form of a spectrum, which is merely a plot of the relative density of each component as a function of frequency. It is important to distinguish this kind of spectrum from one produced by an optical or other spectrometric instrument so often used in chemistry. For clarity, we will refer to the spectrum produced upon Fourier transformation as the Fourier spectrum or the Fourier domain signal.

A number of the basic properties of Fourier transforms and their relation to spectroscopic measurements have been discussed by Horlick.[11] In the next section, the important role of Fourier transforms in implementing correlation-based signal processing operations will be emphasized.

2. Correlation-Based Signal-Processing Operations

One of the most common areas where correlation techniques are utilized in chemistry is in the software processing of signals. Here processing refers to such varied operations as smoothing, differentiation, resolution enhancement, deconvolution, and signal detection. In the main, these operations are applied to a signal after it has been digitized and stored in a computer, and as such, this type of signal processing is often referred to as digital filtering. Digital filtering techniques have been used for many years to process chemical signals. The basic equation for digital filtering is the correlation equation, equation (1). In a now classic paper, Savitzky and Golay[12] presented several types of correlation filters for smoothing and differentiation of data. In recent years, the Fourier domain route of correlation [equations (6) and (7)] has become popular for implementing digital filtering operations on a variety of chemical signals.[13-19]

In this chapter, the application of Fourier domain digital filters to signals of chemical interest will be presented. In particular, a digital filter based on a simple trapezoid is shown to be uniquely versatile. When applied to signals using equation (7), smoothing, differentiation, resolution enhancement, or deconvolution can easily be implemented. In addition, digital filtering is often used as a preprocessing step in signal detection operations. This aspect of correlation will also be discussed.

2.1. Basics of Fourier Domain Digital Filtering

The Fourier transform route of correlation is an effective way of implementing digital filtering operations, and all examples illustrated in this section were carried out in this fashion. A fast Fourier transform (FFT)[20] was used to carry out the transformation of the signals and the inverse transformation of the filtered Fourier domain signal. This algorithm, of which several versions are available, is simply a rapid, efficient means for calculation of the Fourier transformation of a set of points. The input to a typical program is a set of real data (i.e., a digitized waveform in a computer). The output of a typical FFT program consists of two series, the real part of the transform $[X(J)]$ and the imaginary part $[Y(J)]$. These two outputs can be used to generate two additional series, the amplitude Fourier spectrum of the original waveform and the phase spectrum of these Fourier frequencies. The amplitudes of the Fourier frequencies $[A(J)]$ are calculated from the real and imaginary outputs by taking the root sum of squares of the two series, i.e.,

$$A(J) = [X(J)^2 + Y(J)^2]^{1/2} \tag{10}$$

The phases of these Fourier frequencies $[P(J)]$ are calculated using the following equation:

$$P(J) = \arctan [Y(J)/X(J)] \tag{11}$$

All these outputs are illustrated in Figure 12. The original input waveform (Figure 12a) is a typical peak signal. The real output of the FFT for this input waveform is simply a damped cosine wave. The frequency of this cosine wave depends on the position of the peak with respect to the origin of the input waveform; the functional form of the damping depends

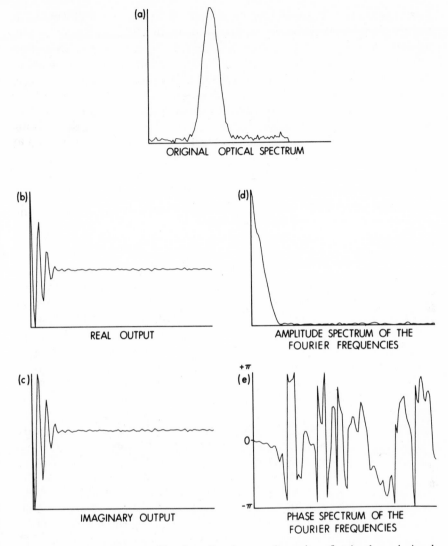

Figure 12. FFT outputs resulting from Fourier transformation of a simple peak signal. The phase spectrum (e) is modulo 2π.

on the shape of the peak in the original waveform.[11] The imaginary output is a damped sine wave with similar characteristics to the real output.

The amplitude spectrum of the Fourier frequencies indicates that the original waveform is composed mainly of low Fourier frequencies along with a relatively large DC level. The amplitudes of the higher Fourier frequencies are small, but their presence is significant in that it is primarily these frequencies that make up the noise in the original waveform. Thus the information about the original peak occurs in a different region of the Fourier frequency spectrum than does some of the noise information. This difference in spectral characteristics enables noise information to be attenuated without loss of signal information. This forms the basis for spectral smoothing operations in the Fourier domain.

The phase spectrum provides information about the phase angle of the individual Fourier frequencies at one specific point on the original waveform. The phase is most useful when calculated with respect to a point on the original wave at which the Fourier frequencies all have approximately the same phase angle. For the waveform illustrated in Figure 12a, this point is the peak maximum. This phase coherence is, in fact, the reason for the existence of the peak when the waveform is interpreted as a Fourier summation of individual frequencies.[8]

It is clear from the phase spectrum that the lower Fourier frequencies which characterize the signal shape are in phase while the higher ones are not. The fluctuation of the phase angle of the higher Fourier frequencies is a definite indication that they arise from noise in the original waveform. In other words, it is unlikely that Fourier frequencies resulting from noise in the original waveform would happen to have the same phase as the Fourier frequencies resulting from the signal (i.e., the peak). Thus, the phase spectrum provides additional information about the distribution of signal and noise frequencies in the Fourier domain.

All filtering operations illustrated in this section were carried out on the real part of the transform $[X(J)]$. Thus, $X(J)$ corresponds to $A(f)$ in equation (7).

A Fourier domain digital filter $[B(f)$ in equation (7)] that is versatile and simple is a trapezoid. It can be characterized by four indices (N1, N2, N3, N4) that define the vertices of the trapezoid.[14,21] Typical Fourier domain digital filters that can be obtained by manipulation of the integer values of the four indices are shown in Figure 13. The vertical dotted line indicates the position of the first point (0 Hz) of the Fourier domain signal. Digital filtering is implemented simply by multiplying the Fourier domain signal by the appropriate filter function. Signal points between N1 and N2, and N3 and N4 are multiplied by the y value of the slope, which varies linearly between 0 and 1. Signal points between N2 and N3 are not altered, and those less than N1 and greater than N4 are set equal to zero. To implement a particular filter the operator simply types in the desired four indices, N1, N2, N3, and N4 at the computer terminal. If N1 is zero and N2 is equal to 1, a filter such as that shown in Figure 13a results; whereas setting N1 negative and N2 positive (>1) results in the filter shown in Figure 13e. The remaining figures indicate other possible filters obtained by varying the values of the indices.

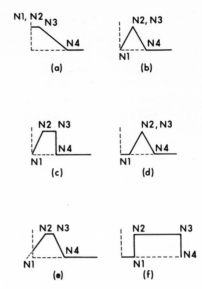

Figure 13. Fourier domain digital filters.

Additional Fourier domain digital filters can be obtained by successive application of the filters shown in Figure 13. Successive application of the filters shown in Figures 13b and 13c results in the filter shown in Figure 14a. Similarly, the filter shown in Figure 14b results from two successive applications of the filter shown in Figure 13e, once with N2 = N3 and the second time with N3 = N4.

Several signal-processing operations can be carried out on spectra using these digital filters. The filter shown in Figure 13a is used for general-purpose smoothing and high-frequency noise elimination. Diagnosis of noise information is often useful and can be carried out using the filter shown in Figure 13f. Differentiation can be accomplished using the filter shown in Figures 13b, 13c, 13d, and 14a, and deconvolution can be approximated using the filters shown in Figures 13e and 14b. These operations are all discussed and illustrated in the next section.

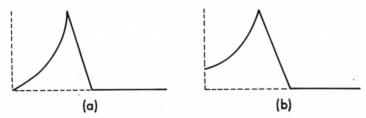

Figure 14. Second derivative (a) and "deconvolution" (b) Fourier domain digital filters.

It should be mentioned here that Fourier domain correlation operations are analogous to the apodization data handling step of Fourier transform spectroscopy.[21] In Fourier transform spectroscopy the measured signal (called an interferogram) is analogous to the real output of the Fourier transformation of a conventional spectral signal (Figure 12b). Apodization (digital filtering) is applied directly to the interferogram before it is transformed to the desired spectrum.

2.2. Smoothing

Smoothing is one of the most frequent software processing operations implemented on chemical signals. The basic aim of smoothing is to improve the signal-to-noise ratio of a signal, thereby enabling a more precise measurement of the desired chemical information. Numerous approaches exist for implementing smoothing, but most are based on a direct application of the correlation integral[12] or its Fourier domain equivalent.[13] It was noted in the last section that some of the noise information in the original waveform appears at different Fourier frequencies than does the signal or peak information, with the noise Fourier frequencies dominating the upper or higher Fourier frequencies. On the basis of this different distribution of signal and noise information among the Fourier frequencies, it is clear that a simple boxcar truncation function can be used as a Fourier domain noise-reduction filter. This is shown in Figure 15 which is a pictorial analog of equations (6) and (7). Note that multiplication by the abrupt truncation function in the Fourier domain is exactly equivalent to direct correlation with the sinc function.

The filtering and diagnosis of high-frequency noise on a signal is illustrated in Figure 16. A spectrum is shown in Figure 16a that contains high-frequency noise, particularly on the peaks. This noise was caused by a faulty power supply in a measurement system. The real part of the Fourier transform of this noisy signal (Figure 16b) clearly reveals the presence of excess high-frequency noise. The Fourier domain signal shown in Figure 16b is 256 points long. Application of the filter shown in Figure 13a with N1 = 0, N2 = 1, N3 = 99 and N4 = 100 (trapezoid indices) results in the retransformed spectrum shown in Figure 16c. Note that most of the high-frequency noise has been removed from the spectrum.

From a diagnostic point of view it may be useful to determine the distribution of the noise in the signal domain. This can be accomplished by using a filter with indices equal to 99, 100, 255, 256 (see Figure 13f). The result of applying such a filter is shown in Figure 16d, which indicates that the noise was concentrated in the region of the spectral peaks.

Further examples of the effectiveness of Fourier domain digital filtering in removing high-frequency noise components are shown in Figure 17.

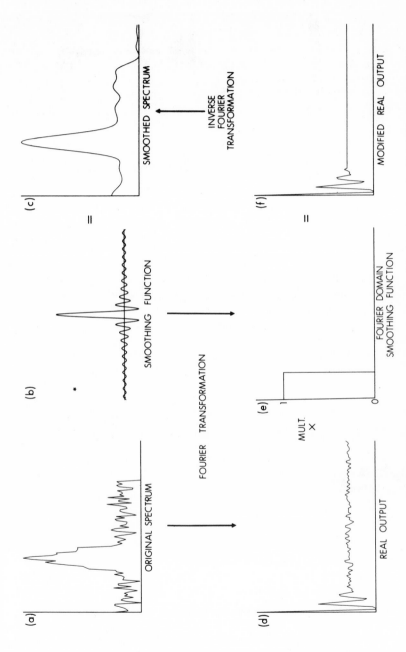

Figure 15. Smoothing of a signal by direct cross-correlation (a,b,c) and by Fourier domain digital filtering (a,d,e,f,c).

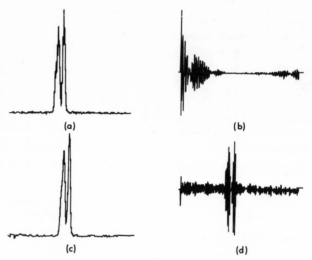

Figure 16. Removal and analysis of high-frequency noise on a signal.

The spectrum shown in Figure 17a contains an intense fixed-frequency noise component. Its transform (Figure 17b) indicates that the noise is concentrated at about the 125th word. A filter with indices equal to 0, 1, 99, and 100 easily removes this narrow-band high-frequency noise component (see Figure 17c). Also, if desired, a software notch filter could be set up for noise of this type.

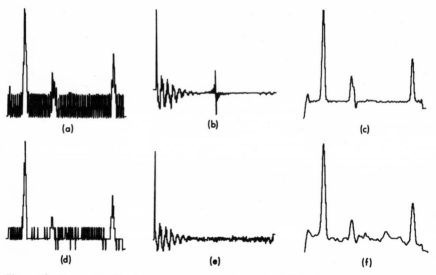

Figure 17. Removal of fixed-frequency noise (a,b,c) and minimization of quantizing noise (d,e,f) by Fourier domain digital filtering.

With low-level signals, quantizing noise can become serious, as shown in Figure 17d. Quantizing noise has considerable high-frequency components as shown by the transform of this signal (Figure 17e). Application of a digital filter with indices equal to 0, 1, 2, and 100 and retransformation yields the spectrum shown in Figure 17f, in which the effects of quantization noise are considerably reduced.

The abrupt filter utilized in Figures 15, 16, and 17 (a–c) may not be the most desirable for certain measurements. If some of the signal information is truncated abruptly, spurious side lobes will result. This is shown in Figure 18a. However, with very noisy spectra it may be desirable to have

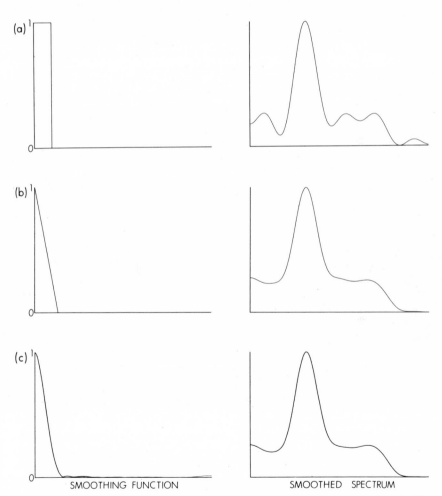

SMOOTHING FUNCTION   SMOOTHED   SPECTRUM

Figure 18. Fourier domain smoothing functions and the resulting smoothed peak. The original peak is shown in Figure 15a.

a low cutoff frequency. This will, in general, necessitate the truncation of some of the higher Fourier frequencies that contribute to the peak information and as such the peak will be broadened. In this case a smoothing function can be used to minimize the side lobes such as a linear truncation (see Figures 18b and 13a). The broadening of the peak is often quite acceptable and is simply a standard trade-off between signal-to-noise ratio and resolution.

Several other smoothing functions can be used such as Gaussians, exponentials, etc., and at this point the question might well be asked, "Is there a smoothing function that will optimize the signal-to-noise ratio for peak height measurement?" Suffice it to say that a considerable amount of work has been reported in this general area, and the characteristics of such a matched filter have been established for optimal peak height measurement[22-25] When the noise is white, the matched filter has the shape of the amplitude spectrum of the Fourier frequencies of the noise-free signal. Such a matched filter for the peak signal shown in Figure 15a is shown in Figure 18c along with the resulting smoothed spectrum. This operation amounts to cross-correlating a noisy line with a noise-free version of itself. This is a powerful correlation technique for peak detection[26] and is further illustrated in Figure 19. In this case, the signal-to-noise ratio is quite low (~2.21), but the cross-correlation function resulting from correlation with a matched filter clearly indicates the presence of the peak signal.

A major point that often bothers many who would like to use digital filtering techniques is the question of signal distortion.[27,28] What must be kept in mind is that often the goal of digital filtering is optimization of the measurement of a *particular* signal parameter. If the desired parameter is peak height, as indicated above, then a so-called matched filter can be used which, although it distorts peak shape by broadening, optimizes the signal-to-noise ratio (peak height/rms baseline noise) for peak height measurement. If it is also desirable to preserve as well as possible the observed line shape, attempt to recover the "real" line shape (deconvolution), or optimally measure other peak parameters such as position, area, or width, then specific digital filters must be designed for this purpose. The *optimal* processing of the signal for all these parameters cannot be simultaneously achieved using a single correlation (filtering) operation.

Clearly a wide range of Fourier domain smoothing functions can be utilized, depending on the signal-processing goals of the experimenter. Thus it is difficult to present hard and fast rules for determining the form and extent of a Fourier domain smoothing function. The choice can be highly dependent on the specific nature of the signal and the noise, the amount of filtering desired and the degree of signal distortion that can be tolerated. In particular, it may be difficult to choose the extent of trunca-

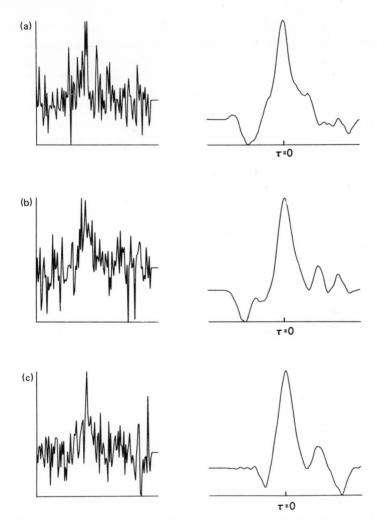

Figure 19. Low signal-to-noise ratio peak signals and the smoothed peaks as determined by cross-correlation with a noise free peak (matched filter).

tion. One simple approach to this problem is to measure, if possible, the signal of interest under conditions that yield a high signal-to-noise ratio. Transformation of the signal will result in a well-defined Fourier spectrum and the point beyond which little or no signal information is present can easily be established.

Bush[16] has presented a quantitative approach to the choice of the truncation point. As noted earlier, the upper or higher Fourier frequency components are essentially due to noise in the original signal. In fact the components are noiselike themselves in appearance and have a character-istic standard deviation. Bush's method of setting a truncation point is to calculate the standard deviation for these upper Fourier frequency com-ponents and then successively repeat the calculation, each time including an additional point from the low-frequency end. When a significant change in the standard deviation is noted, it can be assumed that signal information has been included and, hence, the truncation point can be set.

Some of the problems of truncation and side-lobe generation that occur in Fourier domain smoothing operations do so because of the nature of the signal. This has been discussed in the literature[15] with respect to filtering electroanalytical data and a clever but simple solution presented. The nature of the problem and the solution are presented in Figure 20. The "original" signal is shown in Figure 20a and is an exponential decay. If filtered in the Fourier domain directly (Figures 20b–d), signal distortion results. However, if the original signal is translated and/or rotated so that the initial and final values are zero before being filtered (Figures 20e–i), the introduction of spurious side lobes can be avoided.

2.3. Differentiation

Differentiation of spectra is often used to modify the spectral infor-mation. The first derivative of a spectral peak is used as an aid in exact peak location[29,30] and higher derivatives are used for peak sharpening, i.e., resolution enhancement.[31,32] The derivative theorem of the Fourier transform[33] states that if the imaginary part of the Fourier transformation of a function is multiplied by a linear ramp starting at the origin, the result is the real part of the Fourier transformation of the derivative of the original function. In other words, the imaginary part is multiplied by f. This is summarized by equation (12).

$$
\begin{array}{ccc}
a(t) & & a'(t) \\
\Big\downarrow {\small \begin{array}{l}\text{Fourier}\\\text{transformation}\end{array}} & & \Big\uparrow {\small \begin{array}{l}\text{Inverse}\\\text{Fourier}\\\text{transformation}\end{array}} \\
A(f) \quad \times \quad f & = & f'A(f)
\end{array}
\tag{12}
$$

Multiplication by this linear ramp function in the Fourier domain elim-inates the DC level and attenuates low Fourier frequencies with respect to high Fourier frequencies. It simply amounts to a high-pass digital filter. The accentuation of high Fourier frequencies is a well-known characteristic of a differentiation step.[34] The accentuation should not be carried out beyond the point at which the signal Fourier frequencies disappear or the resulting derivative will be very noisy. Thus, some low-pass digital filtering of the Fourier frequencies should always be used in conjunction with differentiation. This is directly analogous to techniques used in the design

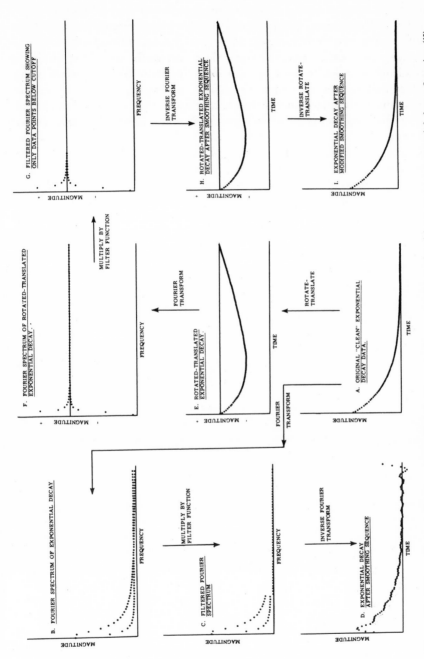

Figure 20. Effects of conventional and modified Fourier transform smoothing algorithms on an exponential decay function.[15]

of analog differentiating circuits using active filters. In addition, as with analog differentiation, the quality of the differential is dependent upon the high- and low-frequency cutoffs relative to the signal frequencies.

Such a Fourier domain digital filter can easily be set up using the trapezoid function as shown in Figure 13b. The effect of such a digital differentiating filter on a spectrum is illustrated in Figure 21. The original spectrum is a doublet and is shown in Figure 21a. The Fourier domain differentiating filter is shown in Figure 21b, and it was applied in a manner analogous to that depicted in Figure 15. Note that both high- and low-pass filtering are readily carried out in one simple multiplication step. The resulting first-derivative spectrum is shown in Figure 21c. The first derivative is perhaps of minimal use in processing peak-type signals, although it has found application in detecting peak locations using its zero crossing. However, if the filter shown in Figures 13b and 21b is applied instead to the *real* part of the Fourier transform of the signal and the resulting

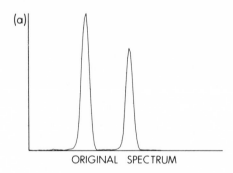

ORIGINAL SPECTRUM

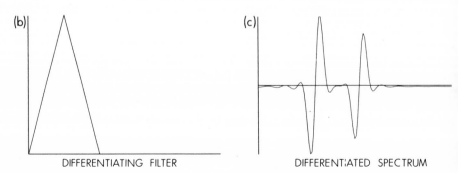

DIFFERENTIATING FILTER   DIFFERENTIATED SPECTRUM

Figure 21. First derivative of a spectrum. The Fourier domain differentiating filter (b) is applied in a manner analogous to that depicted in Figure 15.

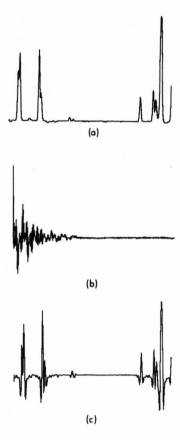

(a)

(b)

Figure 22. Resolution enhancement using a "first-derivative"-type Fourier domain digital filter.

(c)

function (which now is the imaginary part of the Fourier transform of the first derivative of the signal) is retransformed, a resolution-enhanced signal is obtained. This is illustrated in Figure 22, in which the real part of the Fourier transform (Figure 22b) of the spectral signal shown in Figure 22a is multiplied by the digital filter shown in Figure 13b with indices equal to 1, 125, 125, and 250. Note that the signals and transform in Figure 22 are all 512 points long. The resulting resolution-enhanced signal is shown in Figure 22c.

Higher derivatives can easily be obtained by successive application of this filter. A second derivative is obtained by multiplying in the Fourier domain by f^2, i.e.,

$$a(t) \qquad\qquad a''(t)$$
$$\downarrow \text{Fourier transformation} \qquad\qquad \uparrow \text{Inverse Fourier transformation} \qquad (13)$$
$$A(f) \quad \times \quad f^2 \quad = \quad f^2 A(f)$$

The form of the second-derivative Fourier domain filter is shown in Figure 14a. As with the first derivative, low-pass filtering is necessary to control noise. This filter may be achieved by successive application of the filters shown in Figures 13b and 13c. It has also been found that very similar results can be obtained with the filter shown in Figure 13d.[14]

Second differentiation is a powerful technique for resolution enhancement. An example presented by Kelly and Horlick[35] is shown in Figure 23. The signal is a noisy Lorentzian doublet. The composition of the doublet is shown in Figure 23a and the actual noisy signal upon which the second-derivative resolution-enhancement operation was performed is shown in Figure 23b. The result is shown in Figure 23c and indicates the excellent resolution enhancement possible with a relatively simple Fourier domain correlation operation.

2.4. Deconvolution

In the introduction it was mentioned that convolution (equation 5) could be used to describe the effect of an instrument on a signal. It is often desirable to remove, if possible, the effects of such a convolution from a

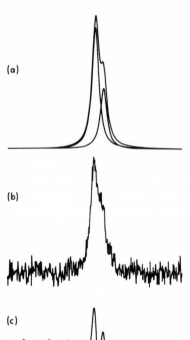

(a)

(b)

(c)

Figure 23. Resolution enhancement using a Fourier domain second-derivative digital filter.

signal. Procedures for reversing the convolution operation are generally referred to as deconvolution. Various deconvolution procedures have been widely used to process signals. An early paper by Bracewell and Roberts[36] dealing with applications to radioastronomy is an excellent introduction to deconvolution methods. Numerous other authors have studied deconvolution methods.[37-46] Stockholm and Cannon[46] present a so-called blind deconvolution, useful in situations where it is either difficult or impossible to know or measure the instrument response function. They have applied their technique to the deconvolution of early Caruso recordings and blurred photographs.

As with correlation, convolution in the Fourier domain can be described as a simple multiplication analogous to equation (7):

$$A(f) \times B(f) = C_{AB}(f) \qquad (7)$$

where $A(f)$, $B(f)$, and $C(f)$ represent the Fourier transforms of the real signal $a(t)$, the instrument response function $b(t)$, and the observed signal $c(t)$. Deconvolution is simply a division operation in the Fourier domain which attempts to reverse the convolution operation. It can be represented as:

$$A°(f) = C_{AB}(f)/B(f) \qquad (14)$$

where $A°(f)$ is the Fourier transform of the deconvolved signal. The complete deconvolution process can be outlined as:

$$
\begin{array}{ccc}
c(t) & & a°(t) \\
\downarrow \text{\small Fourier transformation} & & \uparrow \text{\small Inverse Fourier transformation} \\
C(f) & \times & 1/B(f) = A°(f)
\end{array}
\qquad (15)
$$

where $c(t)$ is the observed signal and $a°(t)$ the deconvolved signal. It is very important to note that the convolution operation is not, in general, reversible, i.e., $a°(t) \neq a(t)$. This point has been discussed in the literature.[36,44,47]

It is clear from equation (15) that deconvolution is simply a standard cross-correlation operation. It can be thought of as a Fourier domain digital filtering operation with a rather specialized filter function which is the reciprocal of the Fourier transform of the instrument response function.

A common application area for deconvolution is spectroscopy. In spectroscopy, the signal is the spectrum which is to be measured, and the instrument response function is the resolution function which is frequently determined by the slit width (Figure 5). In many cases the response function can be determined by measuring the profile of a narrow line.[44] Such a measured line profile and its Fourier transform are shown in Figure 24. Thus the deconvolution Fourier domain digital filter is simply the reciprocal of the function shown in Figure 24b.

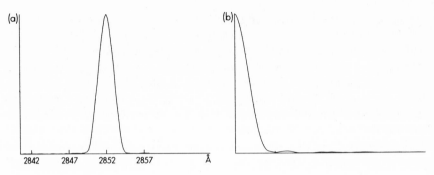

Figure 24. Spectral response function (a) and its Fourier transform (b).

However, the reciprocal of this function becomes very large at higher Fourier frequencies and deconvolution, as with the case of differentiation, can accentuate the high Fourier frequencies to such an extent that noise will obscure the deconvolved signal. The solution to this problem is to combine some low-pass filtering with deconvolution. Thus, a Fourier domain deconvolution filter has a form such as that shown in Figure 25. Note that it is not unlike the second-derivative digital filter. An example of the resolution enhancement that can be achieved with such a controlled deconvolution is shown in Figure 26.[35] The original spectrum was that shown in Figure 23b.

Finally it is possible, using filters such as those shown in Figures 13e and 14b, to approximate deconvolution resolution enhancement techniques. The main value of these filters for resolution enhancement (as compared to differentiation methods) is that with less attenuation of the low Fourier frequencies, there is less generation of negative side lobes.

These examples illustrate the power and ease of Fourier domain digital filtering. With the simple functions shown in this section, a large number of sophisticated signal-processing operations can readily be carried out. In addition, an important aspect of Fourier domain correlation operations is that it is, in most cases, considerably easier to design a digital

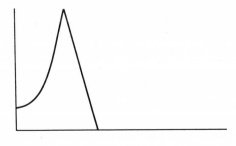

Figure 25. A Fourier domain deconvolution digital filter.

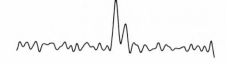

Figure 26. Deconvolution of the noisy Lorentzian doublet shown in Figure 23b.

filter in the Fourier domain than it is in the signal domain. For example, the form of a Fourier domain notch filter is intuitively clear from Figure 17b, while the form of a notch filter that could be directly applied to the signal (Figure 17a) via the convolution equation [equation(1)] is not readily apparent. Thus, the Fourier domain approach to digital filtering considerably enhances an experimenter's capability in developing unique digital filters for specific signal-processing needs.

In conclusion, the comment made about signal distortion towards the end of the smoothing section can be reiterated here. In contrast to smoothing operations, the goal of resolution enhancement is optimization of the measurement of peak position. This is often achieved by derivative and deconvolution techniques such as those illustrated here that generate narrower spectral peak shapes, at times with significant distortion of the peak shape, in order to optimize the precision of peak position measurement. In addition, resolution enhancement is usually achieved at the expense of amplitude signal-to-noise ratio. However, as mentioned before, *optimal* processing of the spectral information for both peak height and peak position cannot be simultaneously achieved using a single filtering operation. Realization of this important concept facilitates intelligent and effective utilization of correlation-based digital filtering techniques.

2.5. Signal Detection

Correlation is often a useful preprocessing step in a signal detection operation.[48] If two signals are composed of peaks, a relatively large value of the correlation function at $\tau = 0$ is a strong indication that the two signals are similar (see Figure 19). In a sense, the correlation operation is a way of concentrating the information about the similarity of two signals into the magnitude of a single point. Even the smoothing aspect of correlation discussed earlier can be looked at from the point of view of signal detection, i.e., finding or detecting a peak in a noisy baseline. For example, correlation techniques have been developed to a high degree of sophistication for the detection of radar signals.[25]

A simple example of the signal detection aspect of correlation[8] is presented in Figure 27. The signal (Figure 27a) is a sequence of binary pulses, and the reference is a triplet sequence (Figure 27b). The cross-correlation function between the two waveforms is shown in Figure 27c.

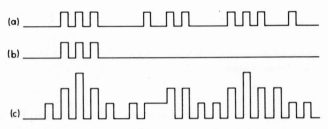

Figure 27. Pattern detection by cross-correlation: (a) signal waveform, (b) sought-for pattern, (c) cross-correlation function.

The maxima in the cross-correlation function indicate positions on the signal where signal information most like the reference waveform is to be found.

The application of cross-correlation techniques to the identification of spectra patterns is shown in Figures 28 and 29.[26] Emission spectra of Co,

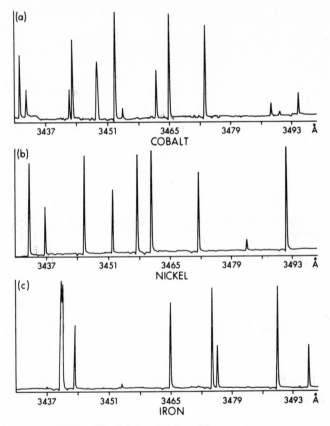

Figure 28. Cobalt, nickel, and iron spectra.

Ni, and Fe in the region of 3430 to 3500 Å are shown in Figure 28. Consider the Co spectrum as the sought-for spectral information. The Co, Ni, and Fe spectra were cross-correlated with the Co spectrum. The resulting cross-correlation functions are shown in Figure 29. Figure 29a is the cross-correlation function of the Co spectrum with itself and in this case it can more accurately be called an autocorrelation function. The pattern has a relatively large maximum at $\tau = 0$, indicating that the two spectra, as expected, have common spectral features. The rest of the function contains information about relative peak separations but, unless the original spectra are quite simple, this information is not readily interpretable. Also, since this pattern represents an autocorrelation, it is symmetrical about $\tau = 0$.

The Co spectrum was then cross-correlated with the Ni spectrum and Fe spectrum. The resulting cross-correlation functions are shown in Figures 29b and 29c. These functions show no distinct maximum at $\tau = 0$, indicating that the two spectra in each case share little similar spectral information. In addition, no symmetry is associated with the cross-correlation function. However, a complex pattern is still present as peaks overlap and coincide as the spectra are shifted relative to each other. All the cross-correlation functions shown in Figure 29 have the same vertical axis of sensitivity, thus, in this example, the behavior of the cross-correlation pattern at $\tau = 0$ allows the identification of the Co spectrum in the measured set of spectra.

It should be noted that many learning-machine approaches to pattern recognition use a similar preprocessing step.[49] The unknown pattern is

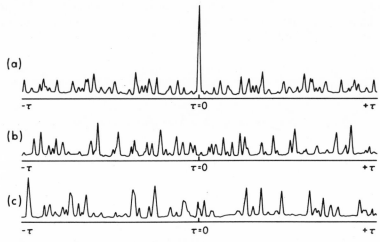

Figure 29. Cross-correlation functions for Co and Co (a), Co and Ni (b), and Co and Fe (c) spectra.

multiplied point by point by a specific vector and then the products are summed to obtain a number upon which a decision can be made about the identification of the pattern.[50] This is, of course, identical to the operation involved in evaluating the magnitude of the $\tau = 0$ point in the cross-correlation function. In addition, an AND-sum operation has been found useful in identifying binary patterns derived from spectra.[51,52] This operation involves evaluating the logical AND-operation between two binary patterns and then summing the resulting logical "ones" and hence is a binary analog of correlation.

3. Correlation-Based Signal-Generation Operations

The correlation operation is ubiquitously present whenever chemical signals are generated or measured. The intimate connection of correlation and convolution makes this clear as any measured signal is the convolution of the "real" signal with the instrument response function. Also, a number of common hardware-based signal-processing techniques that are applicable to the measurement of a rather large number of chemical parameters can be considered as correlation-based techniques. These include the lock-in amplifier and boxcar integrator whose correlation nature has been discussed in some detail in the literature.[6-8]

However, in recent years, a number of measurement techniques have been developed where the correlation operation is more explicitly involved in the signal generation and measurement steps. A number of these measurements will be described in this section to provide the reader with a feel for the nature and utility of correlation-based measurement techniques. Correlation techniques have, of course, been applied to a number of areas other than those highlighted here. The coverage here is not meant to be exhaustive but simply representative. Several additional examples are also presented in Section 4, where work of a more instrumental nature and emphasis is discussed.

3.1. Shot-Noise Autocorrelation Measurement Technique for Photocurrents

The output photocurrent from a photomultiplier tube is not constant but fluctuates due to the quantum nature of light. These fluctuations are generally termed shot noise. In a conventional DC measurement system, the output photocurrent is low-pass filtered to minimize the intensity of the shot noise, and the DC level of the current is taken as a measure of the light intensity. Pao and co-workers[53,54] have indicated that at low light levels and/or wide measurement bandwidths, there may be more signal power in the shot-noise component of the photocurrent than in the DC component. Futhermore, since the amplitude of the shot noise varies as the square root of the signal level, the average value of the square of the shot-noise component (shot-noise power) should be linearly related to light intensity. The measurement of this value can be achieved by autocorrelation of the output shot noise from the PM tube with the value at $\tau = 0$ equal to the desired shot-noise power. A number of workers have successfully used this technique for the measurement of photocurrents.[55-59]

The instrumentation for this autocorrelation measurement is quite simple. All that is needed is an analog multiplier with a low-pass filtered output. A measurement system that has been successfully utilized for this autocorrelation is shown in Figure 30. The analog multiplier unit used to square the signal was a PAR model 230, and it has a low-pass-filtered output. Other analog multipliers such as Analog Devices 426A could also be used.

The linearity of the shot-noise autocorrelation measurement system is shown in Figure 31. The output of the multiplier unit is plotted on a log–log plot vs. the DC photocurrent as measured with a conventional system. The slope of this plot is 45°, indicating that the multiplier output varies linearly with photocurrent and hence is linearly dependent on the incident light level.

3.2. Application of Correlation Techniques in Flame Spectrometry

A number of signal-to-noise ratio enhancement techniques have been applied to flame spectrometric measurements including tuned amplification, lock-in amplification, signal averaging, DC integration, boxcar inte-

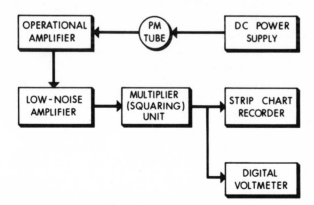

Figure 30. Block diagram of noise autocorrelation measurement system for photocurrents.

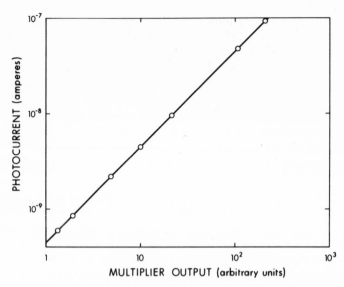

Figure 31. Linearity of the shot-noise autocorrelation measurement system.

gration, and photon counting, some of which, as mentioned above, can be considered as correlation techniques. However, more recently, an explicit cross-correlation technique has been applied to flame spectrometry for the purpose of signal-to-noise ratio enhancement.[9]

The cross-correlation technique was applied to atomic fluorescence spectrometry. The intensity of the primary light source (a rhodium hollow-cathode lamp) was sinusoidally modulated utilizing a programmable power supply. The photocurrent resulting from the atomic fluorescence of rhodium was cross-correlated (model 3721A Hewlett-Packard Correlator) with the sine wave reference used to modulate the lamp intensity. Cross-correlation was compared to lock-in amplification, and the limit of detection for rhodium was 0.16 ppm using the correlator and 0.91 ppm using lock-in amplification. Also, it was shown that the cross-correlation technique was particularly superior in the presence of significant amounts of impulse noise. This is illustrated in Figures 32 and 33.

In this study, impulse noise was simulated by striking a brief spark from a Tesla coil to the grounded photomultiplier housing. In Figure 32 it is seen that large excursions are produced in the output of the lock-in amplifier whenever an impulse (marked by arrows) occurs. The correlator output shown in Figure 33 is considerably less susceptible to impulse noise. The impulse, which still caused deviation in the output signal, can now be simply discarded as being obviously deviant from the otherwise smooth sinusoidally varying cross-correlation function. Both the lock-in amplifier

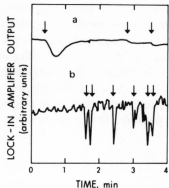

Figure 32. Chart records illustrating the effect of impulse noise on a lock-in amplifier output. Arrows indicate introduction of impulses: (a) 10-s time constant, (b) 1-s time constant.

and correlation measurements were made using a 100-ppm rhodium solution.

Cross-correlation techniques have also been used by Hieftje *et al.*[60] in a study of source modulation waveforms for improved signal-to-noise ratio in atomic absorption spectrometry. In this case sine wave, square wave, ripple, pulse, and pseudo-random pulse modulations of the source intensity were studied. The demodulation step, in order to generate the output signal, can in all cases be interpreted as a single-point cross-correlation with zero delay.[8,61]

3.3. System Characterization using Correlation Methods

There are a number of scientific measurements and studies that necessitate the characterization of the frequency response of a system. Correlation techniques utilizing random input waveforms are beginning to

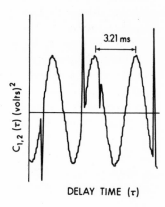

Figure 33. Illustration of the effect of impulse noise on the correlator output. Effective time constant was 13 s.

revolutionize these measurements. In order to understand how correlation techniques enable the characterization of a system, we must digress a bit. Let us imagine the system that we wish to investigate is a "black box" which has an input which enables us to perturb it, and an output which enables us to observe the effect of the perturbation. Our purpose will now be to determine a relationship between the input and output of the "black box" which will enable us to characterize its frequency response. The nature of the frequency response can often then be used to study important physical, biological, or chemical parameters of the system contained in the black box.

One method of characterization that has long been studied and employed by workers in various fields is to determine the *transfer function* of the black box. While the details of the transfer function need not concern us here, one kind of transfer function is particularly useful—the impulse response function. The impulse response function is merely the output that our system would produce if perturbed with an input consisting of an impulse, that is, a very sharp spike.

From Fourier transform considerations[33] it is known that an ideal "infinitely fast" impulse has a flat frequency spectrum, i.e., contains all frequencies. Therefore, perturbing our black box with the impulse is equivalent to simultaneously sending all frequencies into the input at once. Under these conditions, the output of the black box will contain all frequencies, but some will have been attenuated, distorted, phase shifted, or altered in some other way by the system. In short, the impulse response function is effectively a time domain representation of the frequency response of the system. As such, Fourier analysis of the impulse response function is frequently utilized to further aid characterization of the system.

In addition to the impulse response function, the step response function can also be used in a somewhat similar fashion, and this was illustrated in Figure 11 as an example of convolution. Thus the impulse response function is simply the convolution of the input impulse with the system response function. However, the correlation operation can be advantageously employed in a more explicit way for the determination of the impulse response function of a system.

Recall Figure 7, in which the autocorrelation of a random function was shown to produce a sharply peaked function centered at $\tau = 0$. The impulse is therefore merely a phase-related representation of the random waveform. Because of this relationship, perturbing the black box with a random wave should generate an output equivalent in frequency content to that produced by an impulse perturbation. The random waveform, like the impulse, contains all frequencies, although all are not being sent simultaneously into the system. The output of the system also contains all these frequencies altered as described before for the impulse input. It is much more difficult to characterize these alterations, i.e., visually discern the impulse response of the system, in the apparently random output waveform. The key, however, is simply to cross-correlate the output waveform from the system with the random input, which will produce the desired impulse response function.[62] This process is illustrated in Figure 34.

From the foregoing, it is evident that an impulse response function, useful in characterizing a device, can be generated using either an impulse or a random function as an input signal. The important consequence of this observation is that, with a random waveform, the frequency components being used to perturb the device can be spread out over a long period of time, thereby avoiding the sudden, intense perturbation created by an impulse. Because the black box need only accept a small fraction of the total combined amplitude of all frequency components at any given time, the amplitude of each component can be increased many times, and a clearer, more noise-free record of the impulse response function be obtained. Thus the random-input, cross-correlation approach to system characterization is very powerful. Its widespread usage has been somewhat hampered by lack of convenient correlation instrumentation but the appearance of commercial correlators is beginning to make it the method of choice. A wide range of mechanical, electrical, biological, acoustical, and chemical systems have been studied by this technique. As a simple example, consider the black box to be a large motor. In this case the input to the motor would merely be the current to the motor's windings, the output would be an observed rotation of a motor. Information about the relationship between these two quantities would, of course, be of interest to an engineer wishing to employ the motor in a specific application. However, it would be highly undesirable to excite the motor with an impulse, because at the very least, it is likely that the motor's windings would be damaged by this sudden application of a large current. Instead, the engineer can apply a randomly varying current to the motor and observe and correlate the slight variation in output rotation rates which are produced. In this application, an additional advantage derived is that the motor can be observed under normal operating conditions, that is, while it is rotating.

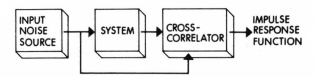

Figure 34. Determination of system impulse response and frequency characteristics using random noise.

Among the more prominent chemical applications of this technique are Fourier transform NMR methods and characterization of electrode processes.

3.3.1. Fourier Transform NMR

It is clear that Fourier transform methods[63] have revolutionized nuclear magnetic resonance spectrometry. It is not our intention here to in any way review this very large field, but simply to indicate how the technique relates to correlation methods. As first implemented,[64] a pulse of radio-frequency energy was used to excite the free induction decay characteristic of nuclear spin relaxation. In effect, the free induction decay signal is merely a special form of impulse response function which, upon Fourier transformation, yields the desired spectrum. More recently, it has been shown by Ernst[65] and others[66] that application of the radio-frequency energy in a random fashion (stochastic excitation) and subsequent cross-correlation can produce even better signal characteristics. In particular, because the instantaneous power of the randomly applied radio-frequency signal can be much lower, the average power can be much higher without producing nuclear saturation. This higher average power in turn provides higher signal-to-noise ratios in the final detected signal or allows reduction in the amount of time necessary to obtain an NMR spectrum at any desired signal-to-noise ratio. A related technique, although not strictly a correlation method, is Fourier synthesized excitation[67] in which the input excitation pulse contains only a selected number of frequencies.

3.3.2. Electrochemical Systems

A number of workers, using techniques similar to those discussed above for Fourier transform NMR, have studied and characterized electrochemical systems. This has included impulse methods,[68] Fourier synthesis techniques,[69-72] and the application of random and pseudo-random noise as applied potential signals.[73]

3.4. Correlation Analysis of Noise—"Fluctuation Spectroscopy"

In recent years, a number of new and unique scientific measurements have been carried out that are based on the correlation analysis of noise. The first measurements, and the majority to date, have involved the analysis of scattered laser light by autocorrelating intensity fluctuations. More recently noise correlation methods have also been applied to the measurement of chemical kinetic information using fluorescence and conductivity fluctuation measurements. This is an exciting and rapidly developing field, and we will briefly review it in this section.

An early review on the subject of utilizing the measurement of light fluctuations as a spectroscopic tool was presented by Wolf in 1964.[74] In particular, it is possible to measure spectral information about the line width of a laser line and/or a Doppler-broadened Rayleigh-scattered laser line by autocorrelating intensity fluctuations. As the technique has evolved, a host of names have become associated with it, among them: light beating spectroscopy, photon correlation spectroscopy, heterodyne laser beat frequency spectroscopy, optical mixing spectroscopy, intensity correlation spectroscopy, intensity fluctuation spectroscopy, and laser Doppler spectroscopy. Reviews of the field have been presented by Cummins and Swinney[75] and Berne and Pecora,[76] and Cummins and Pike[77] have edited a recent volume devoted to photon correlation and light beating spectroscopy.

At present, the largest application of this technique is in the measurement of flow and flow-related parameters such as diffusion coefficients with most effort centered on biochemical systems. A typical experiment involves the measurement of the time-varying scattered radiation (intensity fluctuation) from a flowing system being illuminated by a powerful argon ion laser. Autocorrelation of these intensity fluctuations in the scattered signal can provide information about the flowing system. In one experiment[78] simultaneous measurement of electrophoretic mobility and the diffusion coefficient of bovine serum albumin was carried out utilizing such light-scattering techniques. Since that time, there have been many applications of this technique to flow-related measurements. Among them, a further study by Ware and Flygare[79] applied to fibrinogen, a study of electrokinetic phenomena,[80,81] blood flow,[82] plasmas,[83] air flow,[84] and macromolecular diffusion coefficients.[85] A full review of this rapidly expanding field is beyond the scope of this chapter. However, the interested reader is encouraged to review the original publications and the recent treatment in Volume 2 of this series.[85a]

Noise correlation methods have been extended to the measurement of chemical kinetic parameters utilizing fluorescence and conductivity fluctuations. Fluorescence correlation spectroscopy has been discussed in some detail by Magde et al.,[86] Elson and Magde,[87] and Magde et al.[88] Feher and Weissman[89] have determined the kinetic parameters of the dissociation reaction of beryllium sulfate by correlation of conductivity fluctuations. Chen[90] has also discussed the measurement of kinetic parameters utilizing fluctuations. The basis of this method of determining kinetic parameters can be understood in terms of the random perturbations discussed in the previous section on system characterization.

As chemists, we should recognize that, on a molecular or atomic basis, all species are being continuously perturbed stochastically. Collisions among chemical species, random dissociation reactions, and other occurrences all act as perturbations. The reaction of the chemical system to these perturbations can, however, not be random but must follow the physical and chemical kinetic laws governing those particular species. Thus the measurement of a particular response function of the chemical system should be able to be achieved without the explicit random input depicted for the black box in the last section. The input would be the naturally occurring random perturbations within and around the chemical system under study. Observation of some property of one or more of the species in the system should then provide information about the temporal behavior of that species. Thus in the laser scattering experiments, the intensity fluctuations in the scattered laser light reflect concentration fluctuations due to diffusion of macromolecules.

In the more recent work of Feher and Weissman[89] and Elson and Magde[87] concentration fluctuations are reflected in conductivity and fluorescence intensity fluctuations from which, by autocorrelation, chemical kinetic information can be obtained. To understand these experiments, consider the simple reaction shown below:

$$A + B \rightleftharpoons C$$

We would ordinarily think of this chemical reaction as being at equilibrium. Of course, any true chemical "equilibrium" is really a dynamic system, involving both forward (k_1) and reverse (k_{-1}) reactions. Therefore, the concentrations of all reactants and products in the reaction are continuously changing because of natural, stochastic perturbation around and within them. However, as in the preceding examples, the chemical species cannot react instantaneously to these perturbations but must change in concentration in accordance with their own kinetic characteristics. Thus, if the concentration of species C were observed over a period of time, its kinetic behavior could be deduced by autocorrelation analysis of its concentration fluctuations.

Although this method of measuring chemical kinetics has not been broadly explored and is not yet widely recognized, it is certain to be of great significance in the future. It enables reaction kinetics to be observed under essentially equilibrium conditions, thereby simplifying mechanistic interpretation of the kinetic information. In addition, it is unnecessary to externally perturb the system. Extremely rapid kinetics can be observed; in fact, the speed with which kinetics can be observed is limited only by the response time of the devices being used to monitor the chosen species. These and other potential advantages are discussed further in the references cited in this section. Certainly one "advantage," mentioned by Feher and Weissman, is "the personal satisfaction of using rather than fighting noise." Clearly noise correlation represents a new, powerful, and elegant technique. It provides both unique measurement capabilities for chemical parameters and fresh insight into chemical systems.

Finally a related area, that of calculating molecular motion information by Fourier transforming infrared and Raman spectral band shapes, should be mentioned. The result is a so-called time correlation function that can provide information about molecular rotation.[91,92] In a sense, the laser scattering experiments are just line-broadening measurements. However, the spectral resolution necessary for its observation is beyond the capability of essentially all conventional techniques except perhaps the Fabry-Perot interferometer, and thus new approaches such as the intensity correlation method were developed. In the infrared, however, the broadening of the spectral lines by molecular rotation can be measured directly by conventional spectroscopic techniques and the peak shape transformed by software to give the autocorrelation function. However, more recently, Rayleigh light-scattering techniques have been used to measure orientational correlation functions in liquids as very short time behavior can be observed.[93]

3.5. Correlation Spectroscopy and Interferometry

Correlation spectroscopy is basically a hardware method of cross-correlating a measured spectrum with a stored replica of the sought-for spectrum.[94,95] Typically a mask replica of the sought-for spectrum is positioned in the exit focal plane of a spectrometer and either the mask or the spectrum is oscillated to generate a chopped signal whose amplitude is a maximum when the spectrum most closely matches the exit mask. Note that this is essentially identical to the software cross-correlation operation depicted in Section 2.5 for the Co, Ni, and Fe spectra, except that only the value of the cross-correlation function at $\tau = 0$ is being evaluated with this mask technique. One of the main application areas for this technique has been the detection of gaseous air pollutants and recent studies utilizing this technique have been presented by Strojek et al.[96] and by Walter and Flanigan.[97]

Similar correlation techniques have also been used to process interferograms obtained from a Michelson interferometer as used in Fourier transform spectroscopy.[95,98] With this technique quantitative information can be obtained from an interferogram without the necessity of Fourier transformation. This method has excellent potential for providing simple systems to process interferograms.

3.6. Correlation Chromatography

In normal chromatographic operations, there can be an appreciable delay between the injection of a sample and readout of the peak information. Correlation techniques have been investigated with the idea of achieving continuous chromatographic analysis.[99] As mentioned in Section 3.3, a system can be characterized by measuring its impulse response function. The injection of a sample into a chromatograph can be considered as an impulsive input, thus a normal chromatogram is the impulse response function of the chromatograph. It was seen in Section 3.3 that the same information could be measured utilizing a random input function and cross-correlating the system output with the random input. This is the basis of correlation chromatography.

To produce a correlation chromatogram, a random or pseudo-random command signal is used to control the sample valve to switch it back and forth between the sample stream and carrier gas or standard sample. When the output signal of the chromatograph is cross-correlated with the pseudo-random input control signal, the result is a correlation function that has the appearance of a normal single-injection chromatogram.

Although quite similar to a normal single-impulse chromatogram, the correlogram differs in the following respects. Since it contains the average information obtained from a number of injections, it provides a more reliable estimate of the signal. Also, as new information appears in the output, the oldest is rejected and a new updated correlogram is calculated and displayed. This implies the possibility of creating a chromatographic system in which the sample valve rapidly and repeatedly injects samples and which displays a chromatogram following sample concentration variations more or less continuously. Finally, the correlogram has an increased signal-to-noise ratio compared to a single-impulse chromatogram due to the fact that random noise does not correlate and therefore is automatically cancelled in the operation. Although some of these ideas have been verified experimentally, correlation chromatography suffers certain limitations because of nonlinearities in the chromatographic system. The practical aspects and limitations of correlation chromatography are discussed in detail by Annino and Bullock.[99]

3.7. A Correlation Method for the Measurement of Decaying Exponentials

In a recent paper by Miller et al.,[100] a cross-correlation method for the measurement of decaying exponentials is presented. The measurement system was actually applied to semiconductor transient signals, but it should be applicable to any situation where similar signals are measured. This is an excellent paper to read in order to get a feel for the correlation approach to measurement.

The basic idea of this correlation measurement system is shown in Figure 35. Clearly, this is just a hardware implementation of matched

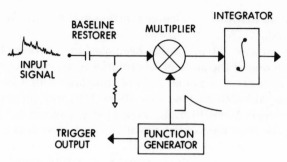

Figure 35. Block diagram of a correlation measurement system for the measurement of decaying exponentials.[100]

filtering (Section 2.2) where the noisy exponential is cross-correlated with a noise-free exponential. The performance of the correlator is discussed and compared with boxcar integration and lock-in amplification for the same measurement. In addition, the utility of alternative weighting functions to the exponential is assessed.

4. Correlation Instrumentation

From the foregoing sections, it is clear that correlation techniques can be effectively applied to a wide range of chemical measurement and signal processing situations. However, one of the most serious drawbacks to the widespread utilization of correlation techniques has been the unavailability of suitable computational methods or instrumentation for carrying out the correlation operation. In the last few years, significant progress has been made in both software and hardware approaches for the implementation of correlation. In addition, present developments in certain large-scale integrated circuits such as diode arrays and CCDs should provide, in the near future, very inexpensive devices capable of sophisticated correlation operations. The areas to which correlation techniques are now being applied have been greatly expanded by these instrumental developments.

4.1. Software Approaches

With the widespread use of small computers in the chemical laboratory, correlation techniques can easily be implemented on a variety of signals using software. Data can be read into the computer in a variety of ways, but for most flexibility in a laboratory environment the basic computer system should include an analog-to-digital converter for acquisition of signals and standard analog outputs such as a plotter and a scope or graphics unit so that the correlation function and signals can be conveniently displayed.

A minicomputer system that has been used for a variety of correlation applications in one of the authors' laboratories is shown in Figure 36. The analog input consists of a sample and hold amplifier with a 50-ns aperture time (Analogic MP-250) coupled to a 10-bit ADC with a conversion time of 10 μs (Analogic M 2810). Data acquisition is initiated by a start pulse indicative of the beginning of the signal and is clocked in at a rate appropriate to accurate sampling of the specific signal.[101] Data can be displayed on an oscilloscope or a strip-chart recorder via a DAC, listed on the DEC writer terminal or permanently stored on magnetic tape (DEC tape) for further processing. The computer is a DEC PDP8/e with 16K of core and a DEC tape-based OS/8 operating system. The software was written primarily in FORTRAN. Machine language (SABR) was used to input data from the ADC and output data to the DAC via a DEC DR8-EA 12-channel buffered digital I/O. The machine language commands could be inserted directly in the FORTRAN program, a feature that has proven to be exceptionally convenient and powerful in our laboratories. Obviously, a

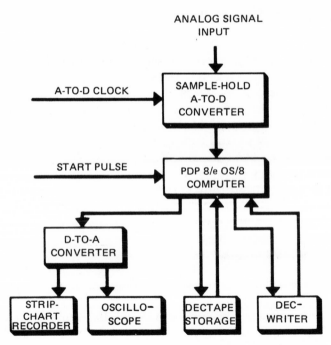

Figure 36. Block diagram of a minicomputer data acquisition system applicable to laboratory correlation measurements.

number of other systems could be used, but that shown in Figure 36 is typical of the capability required in a flexible system for laboratory-oriented correlation applications utilizing a minicomputer.

The basic equation of a discrete correlation operation (equation 2) is easy to program and simple programs for the direct implementation of correlation filtering operations were presented by Savitzky and Golay.[12] However, for moderately large arrays correlation can be quite time-consuming on computers. The Fourier transform route of correlation is often a much quicker computational route. In particular, the fast Fourier transform algorithm has essentially revolutionized the extent to which correlation techniques are utilized for processing scientific signals. The Fourier transform route of correlation was discussed in Section 1 and emphasized throughout Section 2.

Finally, it should be mentioned that many manufacturers are now providing complete software-based systems that are applicable to correlation analysis. Among these are the DECLAB-11/10 systems and the TEKTRONIX digital processing oscilloscope.[102] These systems can be obtained with the ADC, DAC, and graphics hardware necessary for a very powerful correlation signal-processing system. In addition, complete software packages are available such as DEC's Lab Applications-11[103] that contains FFT capability. These and similar systems can provide the experimenter with very flexible and powerful correlation signal-processing capability.

4.2. Hardware Correlators

At the present time, it is really in the hardware area that some of the most exciting advances are occurring in the development of systems that are capable of evaluating correlation functions. Despite the relatively high cost of most commercial hardware correlators, a number of electronic components and subsystems are now available which allow the design and construction of sophisticated but inexpensive correlation instrumentation.

There are, of course, many approaches that can be taken in the design of a hardware correlator. One of the simplest basic designs is that shown in Figure 37a. This correlator consists of a four-quadrant multiplier and a low-pass filter. A four-quadrant multiplier is simply an analog multiplier capable of multiplying two analog signals of any polarity. Units such as Analog Devices model 426A have better than 1% accuracy and sell for less than $50. Units with better than 4% accuracy can be purchased for about $10. Thus the correlator illustrated in Figure 37a is indeed inexpensive.

A remarkably large number of correlation measurements can be carried out with this simple instrument. As illustrated in Figure 10, this correlator forms the basis of the lock-in amplifier. With relatively simple additional circuitry[8,104] inexpensive lock-in amplifiers can easily be con-

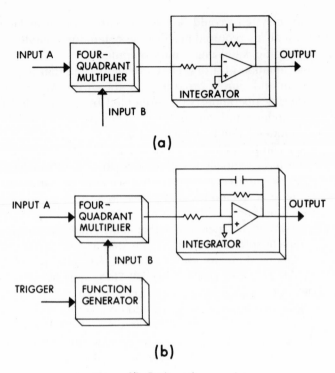

(a)

(b)

Figure 37. Basic analog correlators.

structed from this basic unit. In addition, the correlation discussed in Section 3.1 and the demodulation correlation measurement[60] mentioned in Section 3.2 can both be carried out with this simple correlator.

The versatility and applicability of this correlator can be greatly extended by the addition of a function generator as shown in Figure 37b. This correlator is primarily applicable to repetitive signals. A simple function generator is shown in Figure 38a along with the appropriate waveforms. It consists of a variable delay monostable and a gate monostable, where the gate function is the waveform sent to the multiplier. With this function generator, our correlator becomes the familiar boxcar integrator,[6-8] and in this application the four-quadrant multiplier can even be replaced by an analog FET gate.

A simple modification of this function generator is shown in Figure 38b in which an OA differentiator (high-pass filter) and a clipping circuit are connected to the output of the gate monostable. Now the function generator is capable of generating a variety of decaying exponentials and matched filter correlation measurements can be carried out such as that discussed in Section 3.7 (see Figure 35). Obviously, this approach is applicable to a wide variety of measurements, limited only by the experimenter's ability and ingenuity in designing appropriate function generators.

The correlators illustrated in Figure 37 are limited in that essentially only the value of the correlation at $\tau = 0$ is evaluated, although some variation in τ is possible with the function generators shown in Figure 38 by adjusting the delay monostable. However, for more general-purpose correlation operations, an instrument such as that shown in Figure 39a is necessary.[8]

This correlator is capable of correlating a reference waveform stored in a digital memory with a repetitive analog waveform with the resulting correlation function being plotted on a recorder. The correlator consists of two basic sections, a digital storage and shifting section for the reference waveform and the analog correlation unit previously described. The desired reference waveform is first read into the N-word circulating shift

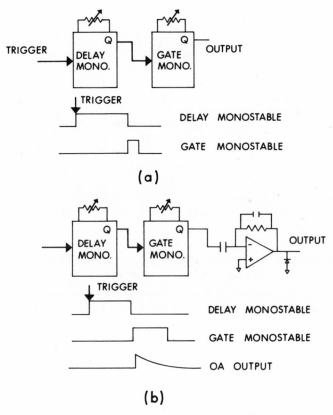

Figure 38. Typical function generators for (a) a gate waveform and (b) a decaying exponential.

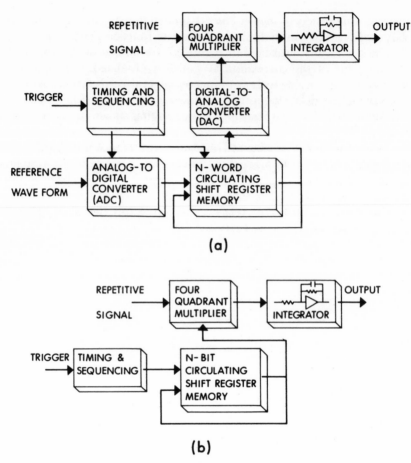

Figure 39. (a) Correlator for repetitive signals. (b) Binary pattern correlator (digital scanning boxcar integrator).

register memory via the analog-to-digital converter. One point of the correlation function between this reference waveform and a repetitive analog signal can then be evaluated by triggering the circulation of the memory with a pulse indicative of the start of the signal repetition. After sufficient products have been integrated to reach a steady-state value at the output of the low-pass filter, the sequencer sends a single extra clock pulse to the shift register which shifts the position of the reference waveform, relative to the start pulse, one clock period. The value of the correlation function at this new relative position is then evaluated by repetitive multiplication and integration and then the shift pulse is again applied. In this manner, the complete correlation function can slowly be evaluated and plotted on a recorder.

Even though this correlator is considerably more sophisticated than those shown in Figure 37, it is not particularly difficult or expensive to build. For example, typical medium-speed 10-bit DACs and ADCs can be obtained for $50 and $100. In the past, a major stumbling block in building such a correlator would have been the memory. However, a wide variety of inexpensive shift register memories are now available and the construction of an N-word circulating shift register memory is relatively simple using these ICs. For example, 1024 by one bit shift registers can be obtained in a single 8-pin IC package (Signetics 2533) for less than $15. Ten of these, appropriately interconnected, provide a 1024-word 10-bit shift register memory. Other memories of shorter lengths are available that combine more than one register on each IC, such as dual–256 bit (Signetics 2527) and hex–32 bit (Signetics 2518) shift register memories. Also the combination of the ADC, shift register memory, and DAC constitute what is known as a transient recorder.[8] These instruments are available commercially (Biomation, 10411 Bubb Road, Cupertino, California 95014) or can be built following designs such as that of Korte and Denton[105] and Daum and Zamie.[106]

Many variations and extensions of the basic design shown in Figure 39a are possible. If it is only necessary to cross-correlate the signal with a binary pattern, this correlator can be simplified to that shown in Figure 39b where the N-word memory is replaced with an N-bit circulating shift register memory. Now a single 8-pin IC can store up to a 1024-bit binary pattern. Such a unit can be useful for pattern detection[52] or can be used as a versatile scanning boxcar integrator.[8] If the signal is not repetitive or if autocorrelations are to be evaluated, then a second memory section must

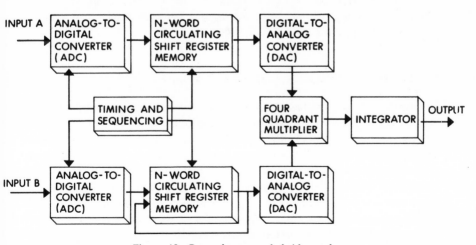

Figure 40. General purpose hybrid correlator.

be added as shown in Figure 40. Also, if desired, the correlation function evaluation could be carried out digitally rather than using the analog approach presented in these examples. At this point, one reaches the rather complex and expensive (but powerful) completely digital hardware correlators available from such sources as Hewlett-Packard,[62] Saicor (Honeywell Signal Analysis Operation, Test Instruments Division, 595 Old Willets Path, Hauppauge, New York 11787), Spectral Dynamics Corporation, P.O. Box 671, San Diego, California 92112, and Elsytic, 212 Michael Drive, Syosset, New York 11791. However, it should be clear from this discussion that correlation instrumentation need not be looked upon as complex and expensive. The simple correlators depicted here are quite capable of carrying out a wide variety of sophisticated and useful correlation operations.

4.2.1. Correlators for Fluctuation Spectroscopy

Although conventional autocorrelation instrumentation is applicable to fluctuation spectroscopy measurements, many workers have constructed their own correlators, particularly autocorrelators capable of processing a PM tube signal measured in a photon-counting mode. The correlators are often referred to as digital correlators or photon correlators. Some are solely hardware based and others combine external hardware and conventional minicomputers.

The general nature of the photon autocorrelation measurement has been discussed by Pusey et al.[85] A 1024-channel digital correlator was presented by Ables et al.,[107] and fast digital correlator for weak optical signals by Corti et al.[108] Digital correlators using minicomputers have been developed by Wijnaendts van Resandt[109] and by Gray et al.[110] Also, a number of simple schemes for measuring autocorrelation functions of fluctuating signals are presented by Kam et al.[111]

4.3. Diode Array Correlation Instrumentation

Integrated circuits based on diode arrays and charge coupled devices (CCDs) show promise of providing remarkably sophisticated correlation instrumentation and operations in very compact packages. These devices were initially developed for electronic imaging applications.[112] In this form a photodiode array has facilitated the development of an analog cross-correlation readout system for spectrochemical applications. More recently, diode arrays have also been fabricated as analog shift registers and tapped analog delay lines. The unique applicability of these devices to correlation measurements will be discussed in this section.

Figure 41. Photograph of a 512-element photodiode array.

4.3.1. A Photodiode Array-Based Correlator for Spectrochemical Applications

A photograph of a 512-element self-scanning linear silicon photodiode array is shown in Figure 41. The thin line across the center is the actual array of diodes. They are 0.001 in. high and the elements are on 0.001 in. spacing; thus this array is 0.512 in. long. The array, complete with scanning circuitry is packaged in an 18-pin dual–in-line IC package. A simplified schematic of the complete integrated circuit is shown in Figure 42. Each photodiode is connected to the output line by a FET switch. The FET switches are controlled by a single bit that is shifted through the shift register. Readout is accomplished using two TTL level signals, a start pulse, and a clock. Although the actual circuitry is somewhat more complex than that shown in Figure 42, the start pulse can be thought of as setting the first bit of the shift register, and the clock then cycles the bit through the shift register, reading out the array.

The photodiodes operate in the charge storage mode and hence are

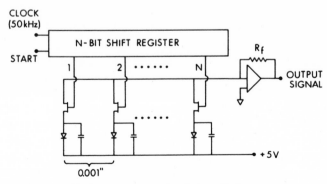

Figure 42. Schematic of photodiode array.

inherently integrating-type detectors. When a particular FET switch is closed by the bit in the shift register, the diode is charged up to its full reverse-bias potential. This reverse-bias charge, stored on the equivalent capacitance of the PN junction, can then be discharged between scans by photon-generated charge carriers (light falling on the diode) and by thermally generated charge carriers (dark current). Thus the signal level necessary on the subsequent scan to reestablish reverse bias on the diode is a measure of the total light intensity and dark current integrated over the time between scans of the array. Thus this device can be considered as a 512-element *analog* shift register with parallel optical inputs and a serial electrical output.

A unique cross-correlation readout system for spectrochemical measurements has been developed utilizing this sensor as the detector.[113] A block diagram of the cross-correlation readout system is shown in Figure 43a. The photodiode array was mounted in the focal plane of the monochromator. A 256-element array was used for this application and about 110 Å of spectral information could be simultaneously observed. A typical

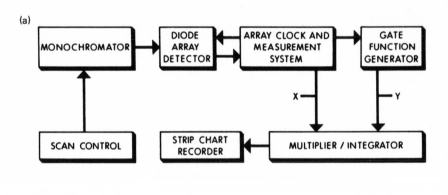

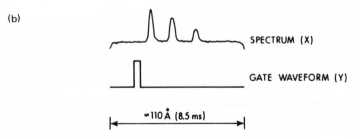

Figure 43. Block diagram of the cross-correlation readout system (a) and schematic waveforms at points X and Y (b).

output signal for a triplet spectrum is shown schematically in Figure 43b
as it would appear on an oscilloscope at point X in Figure 43a. The array
was clocked at a rate of 30 kHz so a complete spectrum (110 Å region) was
readout in 8.5 ms. The repetition time (time between start pulses) was 60
ms (17 spectra/s).

The nature of the correlator is the same as that shown in Figure 37b.
The function generator for this application is shown in Figure 44. It
consists of three delay monostable (DM)–gate monostable (GM) combinations and a system of operational amplifiers. Using this function generator,
any of the output waveforms can be generated in synchrony with the
spectral trace from the photodiode array. The simple pulse waveform is
shown in Figure 43b as it would appear on an oscilloscope at point Y in
Figure 43a.

(a)

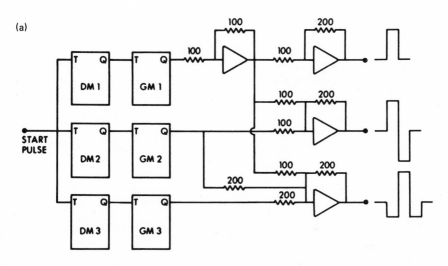

(b)

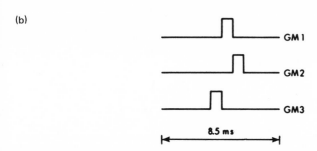

Figure 44. Function generator (a) and schematic representation of output waveforms
from the gate monostables (b) (resistor values are in kΩ).

In order to vary τ and hence cross-correlate a spectral region with a particular pulse waveform, the monochromator was scanned slowly (0.2 Å/ s). This results in the spectrum being slowly scanned past the electronic gate. This is exactly analogous to scanning a dispersed spectrum past a mechanical exit slit (see Figure 5). Thus the pulse gate amounts to an "electronic exit slit." In contrast to the mechanical slit, the electronic slit can take on several unique forms (Figure 44a). Useful modifications of the spectral information can be carried out by cross-correlation with certain types of bipolar pulses. Two examples are shown in Figure 45. The sequence shown in Figure 45a illustrates a cross-correlation that results in resolution enhancement and that in Figure 45b in an approximate first derivative. The bipolar pulse shown in Figure 45a is a square approximation to a second-derivative response function and that in Figure 45b to a first-derivative response function. Thus these direct cross-correlations are analogous to the Fourier domain filtering discussed in Section 2.3.

Thus this analog cross-correlation system provides a unique and versatile approach to spectrometric readout. A wide range of powerful spectral signal processing operations in addition to those illustrated here, can also be effectively implemented with this system. For example, mask correlation spectroscopy discussed in Section 3.5 could be flexibly implemented with this system as the mask could be replaced by a characteristic electronic gating waveform.

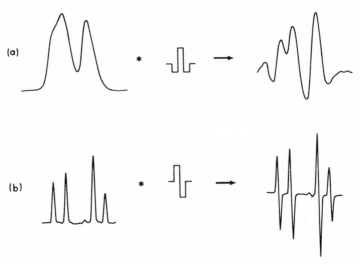

Figure 45. Resolution enhancement (a) and differentiation (b) using the cross-correlation readout system.

4.3.2. Diode Array Analog Shift Registers

As mentioned in the last section, the photodiode array is really an analog shift register with parallel optical inputs and a serial electrical output. Diode array devices are now available that are complete analog shift registers (delay lines) with serial electrical input and output. A simplified schematic of a SAM-128V Serial Analog Memory (128 diodes) available from Reticon Corp. (910 Benicia Ave., Sunnyvale, California 94086) is shown in Figure 46. The diodes are shown only as their equivalent capacitors. The readin shift register controls the sequential sampling of the analog input signal and the readout shift register controls the sequential interrogration of the diode (capacitor) memory elements. The readin and readout rates are independent, thus the analog signal can be reconstructed with a different time base. Also, each memory capacitor has a FET buffer stage, thus readout (unlike that with the photodiode array) is nondestructive. However, with present devices the retention time without significant degradation (because of "dark" current) is 40 ms at room temperature.

Clearly this type of device is potentially an extremely powerful circuit element for incorporation into correlation instrumentation. This *single* IC, with appropriate clocking circuitry, can function as a complete transient recorder. Thus, some of the correlation instrumentation shown in Section 4.2 could be considerably simplified utilizing this device. In addition, this particular serial analog memory possesses some features which make it uniquely applicable to binary sequence correlation. Readout, as mentioned

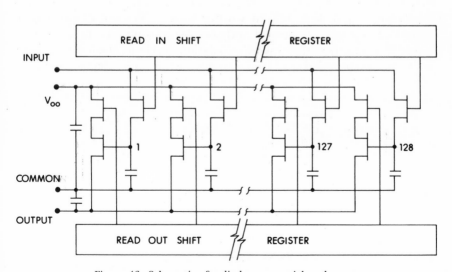

Figure 46. Schematic of a diode array serial analog memory.

above, is controlled by the readout shift register. If a bit exists in the nth stage, the nth cell will be read out. Thus if a specific binary sequence is clocked into the readout shift register, the output will be the summation of the products of the binary sequence and the stored analog information. This application amounts to the rapid cross-correlation of a stored analog signal with a moving binary sequence. Thus a number of powerful signal detection correlations can be carried out with what amounts to a single 18-pin integrated circuit as the correlator.

4.3.3. Tapped Analog Delay Line

A variation of the analog shift register known as a tapped analog delay line is capable of implementing a number of unique correlation operations on analog signals, particularly matched filtering. A schematic of a tapped analog delay line is shown in Figure 47. Basically the device consists of an analog shift register (delay line) with output taps on each storage element. Many applications can be implemented with a small number of delay elements and the unit shown in Figure 47 has only 9 delay elements and taps. The output of the tapped analog delay line is simply the weighted sum (as set by the resistor values) of 9 sequential signal values. Thus an analog signal can be cross-correlated (filtered) by any desired 9-point correlation filter simply by choosing appropriate resistor values. In addition, not all taps need be connected to a single output amplifier, thus an incredibly wide range of filtering operations are possible with such a device.

As a simple example, a conventional low-pass filter can be simulated by giving tap 1 the most weight and the following taps exponentially decreasing weight. It is well known that such a filter can distort a peak signal because only the present and past values contribute to the output value.[12] In the past this was essentially impossible to avoid by real-time analog techniques and could only be avoided by using digital data proc-

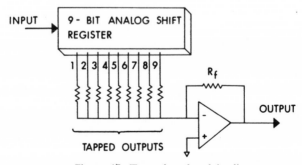

Figure 47. Tapped analog delay line.

essing. However, with the tapped analog delay line filter a trivial modification achieves the desired result. It is simply necessary to give tap 5 the most weight and then give the taps on either side exponentially decreasing weight. In this manner, symmetrical double-sided exponential filtering can be implemented. Clearly, such devices are destined to be widely employed in scientific instrumentation as real-time analog correlators. A serial analog delay line, based on diode arrays, with 24 storage cells and 12 equally spaced taps is presently available from Reticon (TAD-12).

An analog data processor for chart records, whose capability is analogous to that of a tapped analog delay line, has been constructed by Minami et al. [114] This processor was based on an array of ten independent diodes, a detector which is used in conventional punched tape readers. It was capable of evaluating the cross-correlation between a number of 10-point correlation functions [exponential, boxcar (moving average), and Savitzky-Golay type filtering functions] and any type of data in the form of recorder traces.

4.4. Charge Coupled Device Correlation Instrumentation

Charge coupled devices (CCDs) are relatively new electronic devices that are applicable to both electronic imaging and analog shift register applications. A schematic of CCD operation is shown in Figure 48. Basically, the CCD concept involves manipulation of charge on the surface of a semiconductor by moving a potential minimum. [115] In the structure shown in Figure 48, every third electrode is connected to a common conductor. Initially (Figure 48a) voltage V_2 is greater than V_1. This generates a depletion region under electrodes 1, 4, 7, . . . , and any positive charge present or generated near one of these electrodes (such as 1 and 7) will collect near it. If a voltage V_3 ($V_3 > V_2$) is applied to electrodes 2, 5, 8, . . . (Figure 48b), the charge will be transferred from one electrode to the next. In the final sequence (Figure 48c) the voltage V_2 is now applied to electrodes 2, 5, and 8. Thus the CCD is really an elegantly simple analog shift register.

As with diode arrays, most of the initial applications of CCDs have been in electronic imaging, [112] although they are now being applied to both analog and digital shift register applications. A 100-element linear CCD device is available (GEC Semiconductors, East Lane Wembley, Middlesex, England HA9 7PP) that is an imaging device and hence has 100 parallel optical inputs and also has a serial analog input so it can be used as an analog shift register and a line imager either separately or simultaneously. In addition, tapped CCDs are being developed for various analog signal-processing applications. [116,117]

The basic implementation of CCD-based correlation operations is anal-

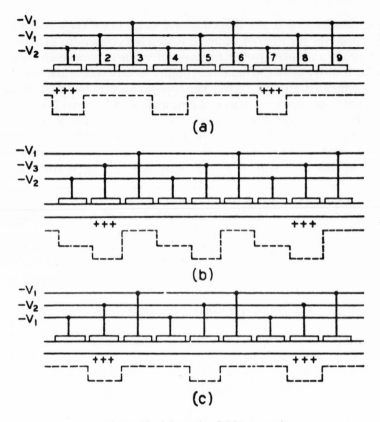

Figure 48. Schematic of CCD operation.

ogous to that with diode arrays. CCDs are somewhat simpler to fabricate and perhaps will become the device of choice in the future. Also, interest remains high in the development of additional solid-state devices with correlation signal-processing capability. Recently, considerable research has been carried out on the development of correlators based on acoustic wave interactions.[118-120]

5. Conclusions

It has been seen in this chapter that correlation techniques are powerful and widely applicable methods for the generation and processing of chemical information. Very often correlation techniques provide unique measurement capability as well as fresh insight into the applicability and limitations of a particular method. Also it was seen that correlation provides a remarkably unifying point of view from which to consider and assess the wide range of measurement and processing techniques available to scientists. Present and future instrumental and conceptual developments are sure to extend even further the power and utility of correlation methods.

6. References

1. Y. W. Lee, T. P. Cheatham, Jr., and J. B. Wiesmer, *Proc. IRE, 38,* 1165 (1950).
2. Y. W. Lee, *Statistical Theory of Communication,* John Wiley & Sons, New York (1960).
3. F. H. Lange, *Correlation Techniques,* D. Van Nostrand, Princeton, New Jersey (1967).
4. W. A. Rosenbleth, *Processing Neuroelectric Data,* The MIT Press, Cambridge, Massachusetts (1959).
5. J. S. Barlow, *IRE Trans. Med. Electron., ME-6,* 179 (1959).
6. G. M. Hieftje, *Anal. Chem., 44*(6), 81A (1972).
7. G. M. Hieftje, *Anal. Chem., 44*(7), 69A (1972).
8. H. V. Malmstadt, C. G. Enke, S. R. Crouch, and G. Horlick, *Optimization of Electronic Measurements,* W. A. Benjamin, Menlo Park, California (1974).
9. G. M. Hieftje, R. I. Bystroff, and R. Lim, *Anal. Chem., 45,* 253 (1973).
10. I. G. McWilliam and H. C. Bolton, *Anal. Chem., 41,* 1755 (1969).
11. G. Horlick, *Anal. Chem., 43*(8), 61A (1971).
12. A. Savitzky and M. J. E. Golay, *Anal. Chem., 36,* 1627 (1964).
13. G. Horlick, *Anal. Chem., 44,* 943 (1972).
14. K. R. Betty and G. Horlick, *Appl. Spectrosc., 30,* 23 (1976).
15. J. W. Hayes, D. E. Glover, D. E. Smith, and M. W. Overton, *Anal. Chem., 45,* 277 (1973).
16. C. A. Bush, *Anal. Chem., 46,* 890 (1974).
17. M. Caprini, S. Cohn-Sfetcu, and A. M. Manof, *IEEE Trans. Audio Electroacoust., AU-18,* 389 (1970).
18. T. Inouye, T. Harper, and N. C. Rasmussen, *Nucl. Instrum. Methods, 67,* 125 (1969).
19. D. W. Kirmse and A. W. Westerberg, *Anal. Chem., 43,* 1035 (1971).
20. R. S. Singleton, *IEEE Trans. Audio Electroacoust., AU-17,* 166 (1969).
21. E. G. Codding and G. Horlick, *Appl. Spectrosc., 27,* 85 (1973).
22. D. C. Champeney, "Fourier Transforms and Their Physical Applications," Academic Press, London (1974).
23. C. W. Helstrom, "Statistical Theory of Signal Detection," Pergamon Press, Oxford (1968).
24. G. L. Turin, *IRE Trans. Information Theory, IT-6,* 311 (1960).
25. W. L. Root, *Proc. IEEE, 58,* 610 (1970).
26. G. Horlick, *Anal. Chem., 45,* 319 (1973).
27. H. P. Yule, *Anal. Chem., 44,* 1245 (1972).
28. H. Tominaga, M. Dojyo, and M. Tanaka, *Nucl. Instrum. Methods, 98,* 69 (1972).
29. J. R. Morrey, *Anal. Chem., 40,* 905 (1968).
30. J. P. Walters and H. V. Malmstadt, *Appl. Spectrosc., 20,* 193 (1966).
31. A. E. Martin, *Spectrochim. Acta, 14,* 97 (1959).
32. L. C. Allen, H. M. Gladney, and S. H. Glarum, *J. Chem. Phys., 40,* 3135 (1964).
33. R. Bracewell, *The Fourier Transform and Its Application,* McGraw-Hill, New York (1965).
34. F. R. Stauffer and H. Sakai, *Appl. Opt., 7,* 61 (1968).
35. P. C. Kelly and G. Horlick, *Anal. Chem., 45,* 2130 (1974).

36. R. Bracewell and J. A. Roberts, *Austr. J. Phys.*, **7**, 616 (1954).
37. P. A. Jansson, R. H. Hunt, and E. K. Plyler, *J. Opt. Soc. Am.*, **58**, 1665 (1968).
38. R. N. Jones, R. Venkataragharan, and J. W. Hopkens, *Spectrochim. Acta*, **23A**, 925 (1967).
39. J. R. Izatt, H. Sakai, and W. S. Benedict, *J. Opt. Soc. Am.*, **59**, 19 (1969).
40. L. Moore, *Br. J. Appl. Phys.*, **1**(2), 237 (1968).
41. B. D. Saksena, K. C. Agarwal, D. R. Pahwa, and M. M. Pradhan, *Spectrochim. Acta*, **24A**, 1981 (1968).
42. T. Vladimeroff, *J. Chromatogr.*, **55**, 175 (1971).
43. J. Szoke, *Chem. Phys. Lett.*, **15**, 404 (1972).
44. G. Horlick, *Appl. Spectrosc.*, **26**, 395 (1972).
45. G. Brouwer and J. A. J. Jansen, *Anal. Chem.*, **45**, 2239 (1973).
46. T. G. Stockham, Jr., T. M. Cannon, and R. B. Ingebretsen, *Proc. IEEE*, **63**, 678 (1975).
47. S. G. Rautian, *Sov. Phys. Usp.*, **66**(1), 245 (1958).
48. W. W. Black, *Nucl. Instrum. Methods*, **71**, 317 (1969).
49. L. A. Powell and G. M. Hieftje, *Anal. Chim. Acta*, **100** (1978).
50. P. J. Jurs, B. R. Kowalski, T. L. Isenhour, and C. N. Reilley, *Anal. Chem.*, **41**, 691 (1969).
51. S. L. Grotch, *Anal. Chem.*, **42**, 1214 (1970).
52. E. G. Codding and G. Horlick, *Appl. Spectrosc.*, **27**, 366 (1973).
53. Y.-H. Pao and J. E. Griffiths, *J. Chem. Phys.*, **46**, 1671 (1967).
54. Y.-H. Pao, R. N. Zitter, and J. E. Griffiths, *J. Opt. Soc. Am.*, **56**, 1133 (1966).
55. J. E. Griffiths, *Appl. Spectrosc.*, **5**, 762 (1967).
56. J. E. Griffiths and Y.-H. Pao, *J. Chem. Phys.*, **46**, 1679 (1967).
57. J. Cleary, *J. Opt. Soc. Am.*, **57**, 841 (1967).
58. R. R. Alfano and N. Ockman, *J. Opt. Soc. Am.*, **58**, 90 (1968).
59. R. Anderson and J. Cleary, *J. Opt. Soc. Am.*, **60**, 531 (1970).
60. G. M. Hieftje, B. E. Holder, A. S. Maddux, Jr., and R. Lim, *Anal. Chem.*, **45**, 238 (1973).
61. P. J. Garforth, *Electron. Inst. Digest*, **6**(8), 7 (1970).
62. Hewlett-Packard Journal, Vol. 21 (November 1969).
63. T. C. Farrar, *Anal. Chem.*, **42**(4), 109A (1970).
64. R. R. Ernst and W. A. Anderson, *Rev. Sci. Instrum.*, **37**, 93 (1966).
65. R. R. Ernst, *J. Magn. Reson.*, **3**, 10 (1970).
66. R. Kaiser, *J. Magn. Reson.*, **3**, 28 (1970).
67. B. L. Tomlinson and H. D. W. Hill, *J. Chem. Phys.*, **59**, 1775 (1973).
68. R. L. Birke, *Anal. Chem.*, **43**, 1253 (1971).
69. H. Kojima and S. Fujiwara, *Bull. Chem. Soc. Jpn.*, **44**, 2158 (1971).
70. S. C. Creason and D. E. Smith, *J. Electroanal. Chem.*, **40**, 1 (1972).
71. S. C. Creason and D. E. Smith, *Anal. Chem.*, **45**, 2401 (1973).
72. D. E. Glover and D. E. Smith, *Anal. Chem.*, **45**, 1869 (1973).
73. S. C. Creason and D. E. Smith, *Electroanal. Chem.*, **36**, App 1 (1972).
74. E. Wolf, *Jpn. J. Appl. Phys.*, **4** (Suppl 1), 1 (1965).
75. H. Z. Cummins and H. L. Swinney, Light beating spectroscopy, in: *Progress in Optics* (E. Wolf, ed.), Vol. 8, p. 135, North-Holland, Amsterdam (1970).
76. B. J. Berne and R. Pecora, *Annu. Rev. Phys. Chem.*, **25**, 233 (1974).
77. H. Z. Cummins and E. R. Pike, (eds.), *Photon Correlation and Light Beating Spectroscopy*, Plenum Press, New York(1974).
78. B. R. Ware and W. H. Flygare, *Chem. Phys. Lett.*, **12**, 81 (1971).
79. B. R. Ware and W. H. Flygare, *J. Colloid Interface Sci.*, **39**, 670 (1972).

80. E. E. Uzgiris, *Opt. Commun.*, *6*, 55 (1972).
81. E. E. Uzgiris, *Rev. Sci. Instrum.*, *45*, 74 (1974).
82. T. Tanaka and G. B. Benedek, *Appl. Opt.*, *14*, 189 (1975).
83. A. Ogata and K. Matsura, *Rev. Sci. Instrum.*, *45*, 1077 (1974).
84. T. S. Durrani and C. A. Greated, *Appl. Opt.*, *14*, 778 (1975).
85. P. N. Pusey, D. E. Koppel, P. W. Schaefer, R. D. Camerini-Otero, and S. H. Koenig, *Biochemistry*, *13*, 952 (1974).
85a. B. A. Smith and B. R. Ware, In: *Contemporary Topics in Analytical and Clinical Chemistry*, Vol. 2 (D. Hercules, G. Hieftje, L. Snyder, and M. Evenson, eds.), p. 29 (1978).
86. D. Magde, E. Elson, and W. W. Webb, *Phys. Rev. Lett.*, *29*, 705 (1972).
87. E. L. Elson and D. Magde, *Biopolymers*, *13*, 1 (1974).
88. D. Magde, E. Elson, and W. W. Webb, *Biopolymers*, *13*, 29 (1974).
89. G. Feher and M. Weissman, *Proc. Natl. Acad. Sci. U.S.A.*, *70*, 870 (1973).
90. Y. Chen, *J. Chem. Phys.*, *59*, 5810 (1973).
91. R. G. Gordon, *J. Chem. Phys.*, *43*, 1307 (1965).
92. H. Shimizu, *J. Chem. Phys.*, *43*, 2453 (1965).
93. J. F. Dill, T. A. Litovitz, and J. A. Bucaro, *J. Chem. Phys.*, *62*, 3839 (1975).
94. D. T. Williams and B. L. Kolitz, *Appl. Opt.*, *7*, 607 (1968).
95. J. H. Davies, *Anal. Chem.*, *42*(6), 101A (1970).
96. J. W. Strojek, D. Yates, and T. Kuwana, *Anal. Chem.*, *47*, 1050 (1975).
97. H. Walter, Jr., and D. Flanigan, *Appl. Opt.*, *14*, 1423 (1975).
98. R. Dick and G. Levy, in: *Aspen International Conference on Fourier Spectroscopy, 1970* (G. A. Vanasse, A. T. Stair, Jr., and D. J. Baker, eds.), Air Force Cambridge Research Laboratories, Special Report, No. 114 (available from the National Technical Information Service) (1971).
99. R. Annino and L. E. Bullock, *Anal. Chem.*, *45*, 1221 (1973).
100. G. L. Miller, J. V. Ramirey, D. A. H. Robinson, *J. Appl. Phys.*, *46*, 2638 (1975).
101. P. C. Kelly and G. Horlick, *Anal. Chem.*, *45*, 518 (1973).
102. R. Ramirez, *Electronics*, *48*(13), 98 (1975).
103. G. Dulaney, *Anal. Chem.*, *47*, 24A (1975).
104. G. Horlick and K. R. Betty, *Anal. Chem.*, *47*, 363 (1975).
105. N. E. Korte and M. B. Denton, *Chem. Instrum.*, *5*(1), 33 (1973).
106. P. H. Daum and P. Zamie, *Anal. Chem.*, *46*, 1347 (1974).
107. J. G. Ables, B. F. C. Cooper, A. J. Hunt, G. G. Morrey, and J. W. Brook, *Rev. Sci. Instrum.*, *46*, 269 (1975).
108. M. Corti, A. DeAgostini, and V. Degiorgio, *Rev. Sci. Instrum.*, *45*, 888 (1974).
109. R. W. Wijnaendts van Resandt, *Rev. Sci. Instrum.*, *45*, 1507 (1974).
110. A. L. Gray, F. R. Hallett, and A. Rae, *J. Phys. E., Sci. Instrum.*, *8*, 501 (1975).
111. Z. Kam, H. B. Shore, and G. Feher, *Rev. Sci. Instrum.*, *46*, 269 (1975).
112. Y. Talmi, *Anal. Chem.*, *47*, 697A (1975).
113. G. Horlick and E. G. Codding, *Anal. Chem.*, *45*, 1749 (1973).
114. S. Minami, S. Tsutsui, and H. Onoda, *Appl. Opt.*, *8*, 1217 (1969).
115. W. S. Boyle and G. E. Smith, *Bell Syst. Tech. J.*, *49*, 587 (1970).
116. J. J. Tiemann, W. E. Engeler, R. D. Baertsch, and D. M. Brown, *Electronics*, *47*(23), 113 (1974).
117. H. F. Beny and C. Husson, *Proc. IEEE*, *63*, 822 (1975).
118. O. Menager and B. Desormiere, *Appl. Phys. Lett.*, *27*, 1 (1975).
119. T. M. Reeder and M. Gilden, *Appl. Phys. Lett.*, *22*, 8 (1973).
120. W. S. Jones, R. A. Kempf, and C. S. Hartmann, *Microwave J.*, *15*(5), 43 (1972).

7. Note Added in Proof

The following papers have been published since this chapter was first written. Although not a comprehensive list, these references and those referred to in these papers will aid the interested reader in keeping abreast of developments in correlation methods. They are grouped according to the chapter headings.

2. Correlation-Based Signal-Processing Operations

1. P. D. Willson and T. H. Edwards, *Appl. Spectrosc. Rev., 12C(1),* 1 (1976).
2. C. G. Enke and T. A. Nieman, *Anal. Chem., 48,* 705A (1976).
3. K. R. Betty and G. Horlick, *Anal. Chem., 49,* 351 (1977).

2.5. Signal Detection

4. G. Beech, *Anal. Chim. Acta, 83,* 133 (1976).
5. T. Hirschfeld, *Appl. Spectrosc., 30,* 67 (1976).

4.2. Hardware Correlators

6. K. R. Betty and G. Horlick, *Appl. Spectrosc., 32,* 31 (1978).

4.3.2. Diode Array Analog Shift Registers

7. G. Horlick, *Anal. Chem., 48,* 783A (1976).
8. K. R. Betty and G. Horlick, *Anal. Chem., 48,* 1899 (1976).

4.3.3. Tapped Analog Delay Line

9. K. R. Betty and G. Horlick, *Anal. Chem., 48,* 2248 (1976).
10. D. L. Doerfler and I. M. Campbell, *Anal. Chem., 50,* 1018 (1978).

5

Signal-to-Noise Ratios in Mass Spectroscopic Ion-Current-Measurement Systems

David W. Peterson and J. M. Hayes

1. Introduction

When an analytical result is based upon a mass spectroscopic measurement, the uncertainty which must be assigned to the result is directly related to the standard deviation of the ion-current measurement. For such an electrical measurement, the standard deviation characterizing the scatter in the observations is correctly termed the *noise*, while the ion-current level is the *signal*. While it is common to consider the precision of chemical analyses in terms of the relative standard deviation ($\equiv$ coefficient of variation), the precision of electrical measurements is always discussed in terms of the inverse quantity, the signal-to-noise ratio:

$$S/N = S/\sigma_S \tag{1}$$

where S represents average signal level and σ_S is the standard deviation of population of individual signal measurements. Any systematic consideration of the precision of mass spectroscopic measurements must be based upon an analysis of the factors ("noise sources") contributing to the denominator in equation (1).

The standard deviation of the population of signal measurements will be affected by all random errors and uncontrolled variations which precede the ion-current measurement in the analytical procedure. Errors in sam-

David W. Peterson and J. M. Hayes • Departments of Chemistry and Geology, Indiana University, Bloomington, Indiana 47401

pling, variations in chemical separations and processing, variations in the efficiency and operation of the mass spectrometer inlet system and ion source, and noise sources within the ion-current-measurement system will all contribute to the observed standard deviation. In this review we will consider only the last category, focusing our full attention on limitations imposed by the ion-current measurement itself.

Our procedure will be to present the expressions which can be used to quantify the relevant noise sources, to summarize the results of experimental observations validating this computational approach, and to conclude by discussing the significance of various factors involved in the selection of an ion-current-measurement system.

2. Noise Sources in Practical Ion-Current-Measurement Systems

The principal components of an ion-current-measurement system are summarized in Figure 1. To provide organization for this discussion, we will treat these components separately and in the sequence in which the signal passes through them. The signal is rendered in final analog form by the signal-conditioning step, and the contribution which any given noise source makes to the final signal variance must be expressed in terms of its effect on this signal. It is adequate in this regard to consider simply two different types of signal conditioning. On the one hand, we shall treat current-follower amplifiers of the type indicated in Figure 2. As noted, the output signal voltage of such an amplifier is directly proportional to the current input, and is given by

$$S_v = -i\bar{G}R \qquad (2)$$

where i is the average ion beam current, $\bar{G}$ is the average current gain of

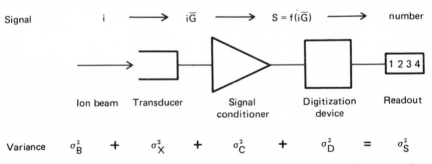

Figure 1. Schematic diagram of an ion-current-measurement system. The signal first takes the form of an ion current, i, experiences a current gain, $\bar{G}$, in the transducer, and is converted into a voltage level in the signal-conditioning step. The over-all signal variance ($\sigma_S{}^2$) observed at the readout is comprised of noise contributions ($\sigma_B{}^2$, $\sigma_X{}^2$, $\sigma_C{}^2$, and $\sigma_D{}^2$) originating in each step of signal processing.

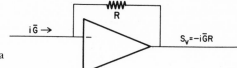

Figure 2. Schematic diagram of a current-follower amplifier.

the transducer, and R is the feedback resistance of the current-follower amplifier. On the other hand, we will consider charge-integrating amplifiers of the type indicated in Figure 3. In this case, the capacitor accumulates charge from the transducer, and the output signal voltage is proportional to the integral of the current. If the current is constant, the output signal is given by

$$S_q = -i\bar{G}t/C \qquad (3)$$

where t is the integration time and C is the capacitance of the feedback capacitor.

A very large variety of mathematical terms must be handled in these discussions, and the nature of the problem condemns us to a profusion of subscripts. After struggling with several "simple" systems, in which we minimized the number of subscripts, we have concluded that "straightforward" and "memorable" are better adjectives than "simple." The resulting system of terms is summarized in Table 1.

As a first step in the prediction of signal-to-noise ratios, we can write

$$\sigma_S{}^2 = \sigma_B{}^2 + \sigma_X{}^2 + \sigma_C{}^2 + \sigma_D{}^2 \qquad (4)$$

and follow with the evaluation of the individual terms. In taking this approach, we are assuming that the noise sources in the various components act independently, an assumption which we have no reason to question.

2.1. Ion Beam Noise, σ_B

If it were somehow possible to construct a perfect ion-current-measurement system, the use of this system for the observation of any real ion beam would result in a "noisy" readout (that is, a constant input ion beam would produce a readout which varied randomly about some constant

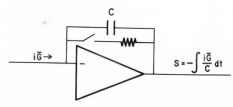

Figure 3. Schematic diagram of an integrating amplifier.

Table 1. Identification of Symbols

A	Pulse-counting efficiency (combined efficiency of transducer and amplifier/discriminator), dimensionless
a	Arbitrary constant
b	Arbitrary constant or Polya parameter [equations (11)–(15)]
C	Capacitance, farads Subscripted by n, nominal
c	Arbitrary constant
d	Signal change corresponding to smallest possible change in digitizer output
e	Electronic charge, 1.6×10^{-19} coulombs; or base of natural logarithms
F	True average count rate, events/second; or frequency corresponding to full-scale input at voltage to frequency converter [equations (43), (62), (63)]
f	Observed average count rate, events/second
Δf	Overall measurement bandwidth, hertz
$\Delta f'$	Bandwidth of analog circuitry, hertz
$G(\ )$	Operator defined by equation (48)
G	Instantaneous current gain of a transducer, dimensionless
$\bar{G}$	Average current gain of a transducer, dimensionless
i	Ion current, amperes
i_1	Leakage current, amperes
i_n	Amplifier current noise density, amperes/$\sqrt{\text{hertz}}$
k	Boltzmann's constant, 1.38×10^{-23} volt coulombs/K°, or arbitrary constant [equations (37)–(39), (58)]
N	True total number of ions or events
n	Number of digitization intervals contributing to a result or number of ions observed in a single observation interval [equations (13)–(15)]
P	Relative standard deviation of an ion current ratio
R	Resistance, ohms; or ion current ratio, dimensionless [equations (50)–(55)] Subscripted by l, leakage; n, nominal
r	Ratio of ion-current-observation times, dimensionless
S/N	Signal-to-noise ratio
S	Signal, volts Subscripted by v, indicating current follower; q, indicating charge integration
T	Temperature, °K or °C, as indicated; subscripted by n, nominal
t	Integration or observation time, seconds
Δt	Interval between digitizer outputs, seconds
V	Applied voltage Subscripted by C, capacitor; fs, full scale; n, nominal
v_n	Amplifier voltage noise density, volts/$\sqrt{\text{hertz}}$
X,Y	Signal-to-noise ratios as defined in connection with equation (64)
α	Temperature coefficient, °C^{-1}; or arbitrary constant [equations (35)–(39)]
β	Voltage coefficient, volts^{-2}; or arbitrary constant [equations (35)–(39)]
δ	Relative difference in ion current ratios, °/oo, as defined by equation (50)
μ	Average gain of a single stage in an electron multiplier
ρ'	Effective deadtime of a counting system, seconds
σ	Standard deviation or noise
σ^2	Variance or noise^2. Noise terms carry up to three subscripts. The first refers to the subsystem in which the noise originates or first appears: B, ion beam; X, transducer; C, signal conditioner; D, digitizer. The second refers to the nature of the output signal: v, current follower; q, charge integration. The third refers to the type of device or to the mechanism responsible for generating the noise: f, Faraday cup; m, electron multiplier, secondary emission noise; j, Johnson noise; t, noise due to ambient temperature variations; a, amplifier noise. Other σ terms refer to the standard deviation in the variable designated by the subscript.
τ	Amplifier time constant, seconds

average value). This irreducible noise originates in the ion beam itself, and is termed "shot noise." The effects of shot noise are colloquially referred to as those of "ion statistics," and arise simply because the arrival of ions at the detector is a random process. Quantitatively, when an average of N ions is collected in each period of observation, the standard deviation of the population of individual observations will be $\sqrt{N}$.[1] Repeated 1-s observations of an ion current of 10^4 ions/s will not furnish a sequence of readouts precisely equivalent to 10^4 ions. Instead, the observations will scatter appreciably ($\sigma = \sqrt{10^4} = 10^2$), with a range of readouts equivalent to 9800–10,200 ions being required to contain 95% of the observations.

Presentation of the shot noise in terms of customary electrical units is straightforward. The relationship between an electrical current and a discrete number of singly charged particles is given by

$$i = Ne/t \tag{5}$$

where i is an electrical current in amperes, e is the electronic charge (1.6 $\times$ 10^{-19} coulombs), and N is the number of particles observed in a time interval t in seconds. The variance in i, the shot noise, is then given by

$$\sigma_i{}^2 = (\partial i/\partial N)^2\sigma_N{}^2 = (e/t)^2 N = ie/t \tag{6}$$

where t, the observation time for the current measurement, is given by $1/2\Delta f$, where Δf is the bandwidth of the measurement system in hertz. In systems employing current-follower amplifiers, t is given by 2τ, where τ is the time constant ($\tau = RC$) of the amplifier. In many cases the bandwidth of the measurement system is not controlled by the amplifier, but by the digitization procedure, with t being determined directly by the integration time employed.

The shot noise is modified by the transducer and signal-conditioning components prior to its observation. The output signal for a system containing a current-follower amplifier is given by equation (2), and the ion beam shot noise then appears as

$$\sigma_{Bv}{}^2 = (\partial S/\partial i)^2\sigma_i{}^2 = (\bar{G}R)^2(ie/t) = 2ie\bar{G}^2R^2\Delta f \tag{7}$$

Alternatively, an ion-current-measurement system might employ an integrating amplifier. In such a case the output signal is given by equation (3), and the ion beam shot noise contribution will be given by

$$\sigma_{Bq}{}^2 = (\partial S/\partial i)^2\sigma_i{}^2 = (\bar{G}t/C)^2(ie/t) = eit\bar{G}^2/C^2 \tag{8}$$

2.2. Transducer Noise, σ_X

The transducer converts the ion beam into an electrical current. A mathematical expression can be used to describe this process:

$$i_{\text{electrical}} = \bar{G}i_{\text{ion beam}} \tag{9}$$

2.2.1. Faraday Cup

The Faraday cup, or collector electrode, is the simplest and most reliable transducer used in mass spectroscopy. As ions strike the Faraday cup, the charge is transferred from the ion to the metal electrode. The Faraday cup has, therefore, a gain of exactly unity. Having no active components, the Faraday cup makes no independent contribution to the total noise in the ion-current-measurement system, and we write

$$\sigma^2_{Xvf} = \sigma^2_{Xqf} = 0 \tag{10}$$

Notwithstanding this fundamental innocence, it remains true that when a Faraday cup is designed or used improperly, spurious results or uncontrolled variations can be observed. On the one hand, ion reflection, a phenomenon in which some ions collide elastically with the electrode and are reflected from the collector without charge transfer,[2,3] can cause the apparent gain to assume values less than unity. This problem can ordinarily be adequately controlled by attention to the geometry of the collector electrode, arranging a deep cup in which chances for reflection and escape are minimized. On the other hand, the uncontrolled loss of secondary electrons from the collector can also cause the apparent gain to assume values greater than unity. This problem is controlled by the use of magnetic and electric fields to ensure retention of secondary electrons in the collector. Finally, improper shielding in the collector system can allow uncontrolled signal variations due to the collection of ions and secondary electrons reflected from other surfaces.

2.2.2. Electron Multiplier

Multiplier detectors have inherent current gains in the range 10^4-10^9, thus greatly increasing the sensitivity of an ion-current-measurement system and, because the higher currents allow the use of lower value feedback resistors, allowing operation at rather high bandwidths. The following discussion is based on the Allen-type multiplier,[4,5] which contains an array of discrete dynodes, but the same principles can be applied to Venetian-blind-type multipliers[6] and to various types of continuous dynode electron multipliers.[7-9]

To consider possibly significant noise contributions arising in the electron multiplier, we can first observe that any variation in the gain of the multiplier will give rise to a variation in the signal. Since such gain variations are known to occur, they must be regarded as a potentially important noise source. A second noise source of potential importance arises in the dark current of the electron multiplier. The dark current itself does not represent a noise source, since it is quite constant and is best treated as a systematic offset rather than a random error. However, the shot noise in the dark current can be considered as a noise source. Examining this possibility, we find that the magnitude of the dark current in a typical Allen-type multiplier is less than 5×10^{-12} A at a gain of 10^{6}[10] and can state, therefore, that shot noise in the dark current will reach an importance equal to that of shot noise in the ion beam only for ion currents of less than 5×10^{-18} A. On this basis we conclude that uncontrolled variations in the gain of an electron multiplier represent the principal noise source requiring detailed consideration. Such gain variations can be grouped under the headings systematic and random.

Systematic variations are observed between ion beams having different translational kinetic energies, different masses, different compositions, and/ or different velocities.[11] These variations originate at the first dynode (or "conversion dynode," the dynode at which the ion current is converted into an electron current) of the multiplier, where ions having different energies (or different masses, etc.) tend to produce different yields of secondary electrons.[11-15] These variations in first-dynode gain are carried through the multiplication process and cause electron multipliers to exhibit substantial discriminatory effects. The gain of the multiplier can also be a function of ion-beam intensity, these fluctuations being due either to loading down of the voltage divider used to establish the interdynode potentials[16] or to vary potential gradients in continuous dynode multipliers.[7]

Even when systematic variations in multiplier gain are avoided or taken into account, microscopic or random fluctuations in the gain will persist and represent a random noise source. The clearest evidence for these variations lies in the observation that a series of identical particles does not produce a series of identical charge pulses at the multiplier anode. Instead, a *distribution* of pulse heights, representing stochastic variations both at the conversion dynode and throughout the secondary electron cascade, is observed.[13,17,18] Quantitative evaluation of the noise due to multiplier gain variations requires consideration of the magnitude of the gain variations.

Dietz has provided the following expression for the relative variance in the gain of an electron multiplier having n stages of multiplication[19]

$$(\sigma_G/G)^2 = b_1 + (1 + b_2)(\mu_1)^{-1} + (1 + b_3)(\mu_1\mu_2)^{-1} + \cdots$$
$$+ (1 + b_j)(\mu_1\mu_2\cdots\mu_{j-1})^{-1} + (\mu_1\mu_2\cdots\mu_n)^{-1} \qquad (11)$$

where the b terms are Polya parameters and the μ terms represent the gains of individual dynodes. A Polya parameter is representative of components of variance associated with nonuniform secondary-electron emission from point to point on the dynode surface and with fluctuations in the secondary-electron emission process itself. These latter effects are caused by variations in the sequences of ion-atom collisions which take place as the projectile ions penetrate the surface.[19a] The Polya parameters take values between 0.0 and 1.0, where a larger value indicates a greater spread in the distribution of secondary electrons emitted per collision. The subscripts designate the stages of multiplication ($j \leq n$), and the variance refers to the population of single-particle pulses. In general, all stages after the conversion dynode are identical, and we can write $\mu_2 = \mu_3 = \mu_j \equiv \mu$ and $b_2 = b_3 = b_j \equiv b$. All terms but the first and last in equation (11) are then members of the geometric series $(1 + b)(\mu_1\mu^{k-1})^{-1}$, where $1 \leq k \leq n - 2$. The terms in this series become diminishingly small, and a useful expression for the sum, independent of n, can be obtained for $k \to \infty$. The last term in equation (11) is also negligible, and an accurate and compact approximation for σ_G^2 takes the form

$$(\sigma_G/G)^2 = b_1 + [\mu(1 + b)/\mu_1(\mu - 1)] \tag{12}$$

The variance in the average gain for samples of n particles is then given by

$$\sigma_{\bar{G}}^2 = \sigma_G^2/n \tag{13}$$

For the case in which an electron multiplier is coupled with a current-follower amplifier having a feedback resistance R, we note that the output signal is given by equation (2) and that the term representing noise due to random variations in the gain of the electron multiplier is given by

$$\sigma_{Xvm}^2 = \left(\frac{\partial S}{\partial \bar{G}}\right)^2 \sigma_{\bar{G}}^2 = \frac{(iR)^2\sigma_G^2}{n} = \frac{ieR^2\bar{G}^2}{t}\left[b_1 + \frac{\mu(1 + b)}{\mu_1(\mu - 1)}\right] \tag{14}$$

when an electron multiplier is coupled with an integrating amplifier having a capacitance C, the output signal is given by equation (3) and the term representing noise due to random variations in the gain of the electron multiplier is given by

$$\sigma_{Xqm}^2 = \left(\frac{\partial S}{\partial \bar{G}}\right)^2 \sigma_{\bar{G}}^2 = \left(\frac{it}{C}\right)^2 \frac{\sigma_G^2}{n} = \frac{eit\bar{G}^2}{C^2}\left[b_1 + \frac{\mu(1 + b)}{\mu_1(\mu - 1)}\right] \tag{15}$$

Consideration of the practical significance of this noise source requires knowledge of typical values for the parameters b and μ. In this regard, we can note that a 16-stage electron multiplier exhibiting a gain of 10^6 will have a gain of 2.37 (the 16th root of 10^6) per stage if all stages have the same gain. An overall gain of 10^6 is also provided by adopting $\mu_1 = 2.0$ and $\mu = 2.4$, quite plausible estimates for singly charged ions of mass ~100 and energy ≤ 5 keV.[20] The values for the Polya parameters are subject to more uncertainty. However, analyses of the distribution in pulse heights from electron multipliers by Dietz[18] and by Prescott[21] indicate $b \sim 0.05$. The first-stage Polya parameter is dependent on many factors, including the type of particle, its energy, and surface of the target. Dietz has measured $b_1 = 0.22$ for $Na_2BO_2^+$ and $b_1 = 0.36$ for Rb^+.[18] These measurements were made with 25-keV ions and an Al_2O_3 target. Typical conditions for electron multipliers used in mass spectroscopy involve particles of lower energy (2–10 keV) striking a Cu–Be dynode like those for the latter stages of multiplication. Taking this into consideration, Schoeller[20] estimated a Polya parameter of 0.1 for the first dynode. Inserting $\mu_1 = 2.0$, $\mu = 2.4$, $b_1 = 0.1$, and $b = 0.05$, we obtain

$$\sigma_{\bar{X}vm}^2 = ieR^2\bar{G}^2/t = 2ieR^2\bar{G}^2\Delta f \qquad (16)$$

and

$$\sigma_{\bar{X}qm}^2 = eit\bar{G}^2/C^2 \qquad (17)$$

thus finding that the noise due to random variations in multiplier gain is approximately equal to shot noise.

2.3. Noise Introduced during Signal Conditioning, σ_C

All current-to-voltage amplifier systems are characterized by one or more intrinsic noise sources which contribute to the signal at the output. Here we will consider noise sources in two categories: (1) those primarily within or dependent upon the nature of the feedback circuit of the amplifier, and (2) those within the amplifier itself.

2.3.1. Current-Follower Amplifier

Potential noise sources in the feedback circuit include Johnson noise in the resistor and the effects of uncontrolled temperature variations on the feedback resistor. Within the amplifier itself, a variety of uncontrolled random electronic fluctuations gives rise to voltage and current noise. Finally, we can consider systematic errors due to current and voltage offsets in the amplifier and arising due to the effects of the sizable temperature and voltage coefficients associated with the high resistances sometimes employed in electrometers.

The signal variance due to Johnson noise in the feedback resistor will be given by[22]

$$\sigma_{Cvj}^2 = 2kTR/t = 4kTR\Delta f \qquad (18)$$

where k is Boltzmann's constant (1.38×10^{-23} volt coulombs/°K) and T is the absolute temperature. For any current-to-voltage amplifier (shunt ammeter, feedback electrometer, vibrating reed electrometer) in which the signal voltage is developed across a resistance R, equation (18) will represent the Johnson noise contribution arising in that resistance. In general, Johnson noise contributions arising in other resistances will be far less significant.

High-value feedback resistors frequently have significant temperature coefficients. When the temperature of the feedback resistance is not controlled and is allowed to vary randomly over the course of a sequence of measurements, noise contributions due to the effects of these temperature variations can be significant. To evaluate this noise term, we begin by noting that the relationship between the actual and nominal values of a resistor is given by

$$R = R_n[1 + \alpha(T - T_n)] \tag{19}$$

where R_n is the nominal resistance at a given temperature T_n, R is the actual resistance at temperature T, and α is the temperature coefficient for the resistor. The variance in the resistance is then related to the variance in the temperature by

$$\sigma_R{}^2 = (\partial R/\partial T)^2\sigma_T{}^2 = (\alpha R_n)^2\sigma_T{}^2 \tag{20}$$

The variance in the output signal (equation 2) due to thermally induced variations in the feedback resistance will then be given by

$$\sigma^2_{Cvt} = (\partial S/\partial R)^2\sigma_R{}^2 = (i\bar{G})^2(\alpha R)^2\sigma_T{}^2 \tag{21}$$

where we have substituted $R \approx R_n$ and where σ_T is the rms variation of the time-averaged temperature of the feedback resistor, this variation being measured at the same frequency as the signal itself.

The noise sources within an amplifier are usually summarized in terms of the current and voltage mean square noises per unit bandwidth, $i_n{}^2$ and $v_n{}^2$, respectively.[23] For a current-follower amplifier with feedback resistance R, the net signal variance due to both these sources is given by

$$\sigma^2_{Cva} = [(i_nR)^2 + v_n{}^2]\Delta f = [(i_nR)^2 + v_n{}^2]/2t \tag{22}$$

The overall signal variance due to noise sources in the signal-conditioning process is then given by

$$\sigma^2_{Cv} = \sigma^2_{Cvj} + \sigma^2_{Cvt} + \sigma^2_{Cva} \tag{23}$$

when a current-follower amplifier is employed for signal conditioning.

A potentially significant but easily avoided systematic error arises due to imperfections within the electrometer. Voltage and current offsets at the input of the amplifier can give rise to the appearance of a constant error signal at the output. This error signal can usually be reduced below the noise level by carefully "trimming" the electrometer.

A systematic error which can be of substantial significance in current-follower amplifiers arises in the nonohmic behavior of high-value resistors ($R > 10^8\ \Omega$). Typically, these resistors do not exhibit a strictly linear voltage–current relationship, especially at relatively high voltages ($V > 20$ volts). Instead, the actual resistance R is related to the nominal resistance R_n by[24]

$$R = R_n \exp(\beta V^2) \tag{24}$$

where β is the voltage coefficient and V is the voltage across the resistor. The resulting error, ΔS_v, in the observed signal is then given by

$$\Delta S_v = i(R_n - R) = iR_n[1 - \exp(\beta V^2)] \tag{25}$$

The magnitude of the voltage coefficient varies among resistors, but a typical value is -8×10^{-6}.[24] For this value, the relative errors are 0.0008%, 0.08%, and 8.0% for 1, 10, and 100 V across the resistor.

It is also appropriate in many cases to consider the effects of temperature variations, discussed above in terms of random error, as representing systematic errors which are subject to correction. In this case we note that the error in the observed signal is given by

$$\Delta S_v = i(R_n - R) = iR_n\alpha(T - T_n) \tag{26}$$

A typical value of α for a 10^{10}-Ω resistor is 1000 ppm/°C.[25] Thus, when two signals measured at different temperatures are compared, the relative error will be 0.01% if the temperatures differ by 0.1°C, and 0.3% if the temperatures differ by 3.0°C. Such effects are far less serious for lower-value resistors. A typical temperature coefficient for resistors in the range below 10^7 Ω is 25 ppm/°C.

2.3.2. Capacitive Integration

It is advantageous in some cases to carry out direct analog integration of the signal from the transducer.[26] A signal conditioner capable of this direct integration is schematically represented in Figure 3, and the resulting output signal is given by equation (3). The absence of large input or feedback resistances in such systems leads to the elimination of Johnson noise and can represent a considerable advantage in some specialized systems. Note, however, that practical systems ($C \geq 10^{-10}$ F) require relatively large ion currents and/or long integration times for the generation of sizable signals. Amplifier noise and the effects of temperature variations represent potentially significant noise sources in such systems, and leakage currents in the charge-collecting capacitor can lead to sizable systematic errors.

The signal variance due to amplifier noise depends on the same parameters discussed in connection with current-follower amplifiers [equation (22)]. Taking the same approach in this case, we write

$$\sigma_{Cqa}^2 = [(i_n t/C)^2 + v_n^2]/2t \tag{27}$$

The value of the charge-collecting capacitor is dependent on temperature, and variations in temperature can therefore affect the observed output signal. The relationship between capacitance and temperature is given by

$$C = C_n[1 + \alpha(T - T_n)] \tag{28}$$

where C is the capacitance at temperature T, C_n is the nominal capacitance at some reference temperature T_n, and α is the temperature coefficient. The variance in the capacitance is then related to the variance in the temperature by

$$\sigma_C{}^2 = (\partial C/\partial T)^2 \sigma_T{}^2 = (\alpha C_n)^2 \sigma_T{}^2 \tag{29}$$

The variance in the output signal [equation (3)] due to thermally induced variations in the capacitance will then be given by

$$\sigma_{Cqt}^2 = (\partial S/\partial C)^2 \sigma_C{}^2 = (i\bar{G}t/C^2)^2 (\alpha C)^2 \sigma_T{}^2 = (\alpha i\bar{G}t/C)^2 \sigma_T{}^2 \tag{30}$$

where we have substituted $C \approx C_n$ and where σ_T is the rms variation of the temperature of the charge-collecting capacitor, this variation being measured at the same time intervals as the output signal.

The overall noise contribution due to signal conditioning is then given by

$$\sigma_{Cq}^2 = \sigma_{Cqa}^2 + \sigma_{Cqt}^2 \tag{31}$$

when capacitive integration is employed.

The same amplifier input offset current which can give rise to a systematic output voltage offset in current follower systems will be integrated in capacitive integration systems. This error ($\Delta S_q = -i_{\text{offset}}t/C$) can be of significance in unfavorable circumstances, and careful attention should be paid to trimming the amplifier in order to reduce the effect of the offset current.

A systematic error inherent in the capacitive integration amplifier is the decrease in signal due to the leakage current of the capacitor. The charge stored on the plates of the capacitor can leak away, usually through the dielectric material. The magnitude of this leakage current, i_1, is

$$i_1 = V_C/R_1 \tag{32}$$

where V_C is the potential difference between the plates of the capacitor and R_1 is the resistance of the leakage path. As the integration time increases, the voltage builds up on the capacitor and the magnitude of the leakage current increases. A general expression for the observed signal in the presence of leakage current is

$$S_q = \int_0^t (i\bar{G} - i_1)/C \, dt \tag{33}$$

In order to examine the integrated effects of charge leakage, we can assume that $i\bar{G}$ is constant and rewrite equation (33) as follows:

$$S_q = \int_0^t [(i\bar{G}/C) - (S_q/R_1 C)] dt \tag{34}$$

In order to simplify this equation, we will make the arbitrary definitions $\alpha = i\bar{G}/C$ and $\beta = 1/R_1C$, writing

$$S_q = \int_0^t (\alpha - \beta S_q)dt \tag{35}$$

Rearranging and substituting $V_c = S_q$, we write

$$\int_0^{S_q} \frac{dV_C}{\alpha - \beta V_C} = \frac{1}{\beta} \int_0^{S_q} \frac{dV_C}{\alpha/\beta - V_C} = \int_0^t dt \tag{36}$$

Evaluating the integrals, we obtain

$$-(1/\beta) \ln [(\alpha/\beta) - S_q] = t + k_1 \tag{37}$$

$$\ln (\alpha/\beta - S_q) = -\beta t - \beta k_1 \tag{38}$$

defining $k_2 = \exp(-k_1\beta)$, we obtain

$$\alpha/\beta - S_q = k_2 \exp(-\beta t) \tag{39}$$

where k_2 can be evaluated from the requirement that $S_q = 0$ at $t = 0$, yielding $k_2 = \alpha/\beta$ and, by substitution

$$S_q = i\bar{G}R_1[1 - \exp(-t/R_1C)] \tag{40}$$

whence it follows that the systematic error due to charge leakage is given by

$$\Delta S_q = S_q' - S_q = (i\bar{G}t/C) - i\bar{G}R_1[1 - \exp(-t/R_1C)] \tag{41}$$

where S_q' is used to denote the signal expected in the absence of charge leakage. Clearly, the magnitude of this error is dependent on the R_1C product, or discharging time constant, of the capacitor. Teflon and polystyrene capacitors, which exhibit low leakage currents, have R_1C products equal to 10^6 s.[27] On the other hand, electrolytic capacitors, which have R_1C products of 10^3 s, leak rapidly and should not be used in current integrators. As shown in Figure 4, the relative error due to leakage current in Teflon and polystyrene capacitors is significantly smaller than in electrolytic capacitors.

2.4. Digitization Noise, σ_D

The process of digitization has significance in this context both with regard to its effect on the integration time t, a variable which has appeared in most of the preceding expressions, and with regard to noise contributions originating in this phase of signal processing. Here we will consider the effects of various analog-to-digital conversion devices and of directly digital measurements, that is, ion counting.

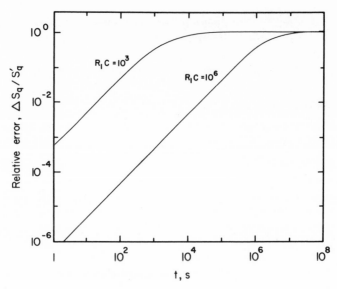

Figure 4. Relative error due to leakage current as a function of integration time in an integrating-amplifier system. See equation (41).

2.4.1. Analog-to-Digital Converters

At the outset, it is useful to consider two aspects of the timing of the digitization process. First, we note that we will consider only processes in which digitization occurs at fixed time intervals Δt, and that during each of these time intervals, the digitizer might observe the signal continuously, as is the case in a device comprised of a voltage-to-frequency converter and a counter, or for only a very brief observation period called the aperture time. We will refer to devices of the first type as integrating digitizers and to devices of the second type as nonintegrating. Unfortunately, the distinction is not always perfectly clear.[28] In many systems, the aperture time is controlled by a sample-and-hold amplifier at the input of the digitization device and can be very brief, occupying only a small fraction of the time interval Δt. Second, we note that the pace of digitization can be either faster or slower than the highest frequencies in the input analog signal. The range of frequencies present at the input is determined by the bandwidth of the analog circuitry, a quantity which we shall denote by $\Delta f'$, distinguishing it from the overall measurement bandwidth, Δf, a variable whose magnitude is often controlled by the digitization process itself. When the pace of digitization is adequate for representation of the highest analog input frequencies, we will have $\Delta t \leq 1/2\Delta f'$, when digitization is slow in comparison to the maximum analog frequencies, $\Delta t > 1/2\Delta f'$.

For the case of a nonintegrating device employed so that $\Delta t \leq 1/2\Delta f'$, we can observe that the effective integration time for a single observation will be controlled by the analog circuitry, with $t = 1/2\Delta f'$. When n digital observations are averaged under these circumstances, the effective integration time will be $n\Delta t$ or $1/2\Delta f'$, whichever is greater, in spite of the fact that the actual signal-observation time might be very brief, being comprised of n aperture times for the digitizer in use. It is also true that when $\Delta t \leq 1/2\Delta f'$, all information present in the analog signal will be effectively transformed into the digital domain, and it will be possible adequately to reconstruct all time-variable aspects of the analog signal from the digital record.

It can happen that a nonintegrating digitizer is applied such that $\Delta t > 1/2\Delta f'$, that is, that digitization occurs slowly in comparison to the highest frequencies presented in the analog input. This is illogical and represents an unsatisfactory situation for a number of reasons. On the one hand, if there is a good reason to provide the analog portion of the measurement system with a high bandwidth, then the same reason must apply equally to the digitizer, since information present at signal frequencies greater than $1/2\Delta t$ is lost during digitization. If, on the other hand, it happens that the sampling rate of the digitizer is adequate for extraction of all useful information from the signal, then we can conclude that the bandwidth of the analog circuitry should be reduced. This conclusion is supported by two facts. First, we note that the effective integration time for each observation is controlled by the analog circuitry and is given by $1/2\Delta f'$. During each Δt interval, only a fraction $(1/2\Delta f'\Delta t)$ of the potential observation time will actually be employed. If n digital observations are averaged, the effective integration time will be $n/2\Delta f'$, which might be appreciably less than $n\Delta t$. Second, we note that high-frequency noise present in the analog signal will be contributing to the instantaneous signal level observed during any given aperture time, and that this noise will contribute to the digital output. This needless incorporation of high-frequency noise in a low-frequency output is termed "aliasing." Since the choice of a relatively long Δt has already indicated that this high-frequency portion of the input signal is uninformative, it would be better to reduce $\Delta f'$ and thus avoid digitizing this noise, at the same time making more efficient use of the informative frequencies in the analog input.

When an integrating digitizer is employed, the effective integration time for a single digital observation will always be given by Δt or $1/2\Delta f'$, whichever is greater, independent of the relationship between $\Delta f'$ and Δt. Because the process of digitization is combined with integration, aliasing is impossible, maximum efficiency is ensured, and the adoption of $\Delta t > 1/2\Delta f'$ is not necessarily inappropriate, although it remains true that information at frequencies greater than $1/2\Delta t$ will be lost.

Independent of the timing considerations discussed above, the process of digitization can affect the variance of the digital output and its relation to the variance of the analog input signal. An important factor in this regard is the resolution of the digitizer, usefully described in terms of the quantization interval, the signal change equivalent to the smallest possible change in the digital output. For analog-to-digital converters which divide the analog continuum into 2^n discrete increments for an n-bit conversion, the quantization interval d is given by

$$d = V_{fs}/2^n \qquad (42)$$

where V_{fs} is the full-scale voltage accepted by the analog input. When analog-to-digital conversion is accomplished by first transforming the signal into the frequency domain, then employing a counter to complete the digitization, the quantization interval is given by

$$d = V_{fs}/F\Delta t \qquad (43)$$

where V_{fs} is the full-scale voltage accepted by the voltage-to-frequency converter, F is the corresponding output frequency, and Δt is the digitization time interval.

An analysis of the generation and/or transmission of noise during the process of digitization depends on whether the noise on the input analog signal is large or small in comparison to the quantization interval. Kelly and Horlick[28] have shown that it is adequate to consider two cases, one in which the input noise is less than $d/2$ and the other in with the input noise is equal to or greater than $d/2$.

An input signal having a noise amplitude which is small in comparison to the quantization interval will yield a constant digital output. The resulting quantization error e_q will be systematic, not random, and will be in the range $-d/2 \le e_q < d/2$ if we assume that the digitizer rounds to the nearest quantization level, or $0 \le e_q < d$ if we assume that the digitizer rounds to the next lower quantization interval. In either case, the range of uncertainty is one full quantization interval.

An input signal having a noise amplitude equal to or greater than $d/2$ will yield a randomly varying digital output. Considering the relationship between the variance of the analog input and that of the digital output, Kelly and Horlick[28] showed that the digitization process added an uncorrelated noise component to the signal. For a single observation, the magnitude of this digitization noise is given by

$$\sigma_D{}^2 = d^2/12 \qquad (44)$$

Repeated digital analog-to-digital conversions will produce a distribution of digital results whose average will not necessarily fall on a discrete quantization step. The variance or quantization error of a signal that has been time averaged is

$$\sigma_D{}^2 = d^2/12n \tag{45}$$

where n is the number of analog to digital conversions.

2.4.2. Ion Counting

Direct digital measurements of ion current levels can be made using an ion-counting system consisting of an electron multiplier, a discriminator, and a counter as shown in Figure 5. An alternative scheme employs a "Daly detector,"[29] in which the secondary electrons produced at a target electrode are accelerated to strike a phosphor, the resulting light pulses being observed with a photon-counting system. In either case, each ion striking the transducer produces a current pulse which is individually amplified and counted. The discriminator has a threshold adjusted high enough to exclude most noise pulses, but low enough that almost all ion pulses are counted. Ions which happen to produce no secondary electrons at the conversion dynode will, of course, be missed; and it is possible for an ion pulse to die out or for the microscopic gain fluctuations to keep the pulse below the discriminator threshold; but, in general, the effects of multiplier gain fluctuations are greatly reduced, and it has been observed that the shot noise of the ion beam [equation (7)] is the only appreciable noise source in the system.[30]

Although ion counting approaches the theoretical limit of precision set by ion statistics, it does have several drawbacks. The most severe restriction is that the dynamic range has an upper limit due to count losses caused by ion coincidence. As a charge pulse travels down the dynode string of the electron multiplier to the discriminator, it obtains a finite temporal width. Whenever two or more ion pulses are separated in time by less than their pulse width, the discriminator will not be able to resolve the pulses, and only the first ion of the multiple event will be recorded. For ion-counting systems and other "paralyzable" (i.e., if struck with an

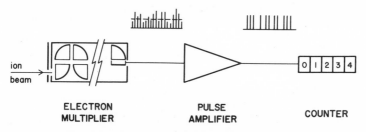

ELECTRON   PULSE
MULTIPLIER   AMPLIFIER   COUNTER

Figure 5. Schematic diagram of ion-counting system.

infinitely fast particle flux, the counter would register no events at all, see reference 32) counting systems, the relationship between the true count rate, F, and the observed count rate, f, is given by[31]

$$f = AFe^{-F\rho'} \tag{46}$$

where A is the efficiency of the multiplier and pulse discriminator/counter and ρ' is the effective deadtime, an experimentally measurable time interval related to the minimum pulse-pair resolving time.[31]

Because the functional relationship between the observed and true count rates is known, it should be possible to apply a correction to the observed count rate, obtaining an accurate value for the true count rate. There are, however, two problems in this regard. First, the efficiency term, A, is usually not well known. In practice, this is of little consequence, since all calculations are ordinarily based on *relative* count rates. In these circumstances, the (unknown) A values can always be cancelled between the two or more count rates being compared. Indeed, the fact that A can be significantly less than unity is seldom mentioned in polite society, and most "correction formulas" [like equation (47) below] simply omit any consideration of the efficiency term. The second problem stems from the form of equation (46), which cannot be directly solved for F in terms of f. The conventional solution to this problem involves the use of any one of a number of formulas, all of which are approximate.[32] For example, the very commonly used correction formula

$$F = f(1 + f\rho') \tag{47}$$

is in error by more than 0.1% for $F\rho' > 0.026$ and by more than 1% for $F\rho' > 0.086$.

Here we would like to point out that numerical methods, exact to within any specified tolerance, can be employed for the solution of equation (46), thus avoiding the approximations incurred in the use of "correction formulas." As a first step in obtaining a numerical solution, it is necessary to define the following function:

$$G(F) = Fe^{-F\rho'} - f \tag{48}$$

Arbitrarily assuming perfect efficiency, we note that $G(F)$ will become equal to zero when the correct value of F is found for a given observed count rate and effective deadtime. The problem, then, is to find the root, $G(F) = 0$, of equation (48). Newton's method[33] finds the required solution iteratively, assigning successive values according to the expression

$$F_{n+1} = F_n - G(F_n)/G'(F_n) = F_n - (F_n e^{-F_n\rho'} - f)/[e^{-F_n\rho'}(1 - F_n\rho')] \tag{49}$$

where, for the first iteration, it is adequate to set $F_1 = f$. When two successive values of F_n are identical, the root has been found and F has been calculated without recourse to approximation. The solution converges rapidly, but it is required that $F\rho' < 0.35$.

Whenever substantial count-loss corrections are made, it is evident that the ion-counting measurement must be regarded as having two parts: (1) the observation of the unknown ion current, and (2) the measurement of the effective deadtime. When assessing the uncertainty in the ion-current measurement, uncertainties originating in *both* these procedures must be considered. While a full consideration of this question is beyond the scope of this review, it should be noted that Hayes and Schoeller[32] have shown that the uncertainty in the effective deadtime assumes critical importance and that the maintenance of a required accuracy of ion-current measurement can have the effect of requiring that count-loss corrections greater than some critical level be completely avoided, thus placing an absolute upper limit on the count rate which can be employed.

2.5. Prediction of Overall System Performance

The expected signal-to-noise ratio for a given system is calculated from equation (1), where the numerator is evaluated using either equation (2) or (3), as appropriate, and the denominator is evaluated by full expansion of equation (4). A single example should suffice.

Consider a system intended for high-speed scanning of the mass spectra of organic molecules. Typical specifications might be as follows:

Ion current range: 10^{-10}–10^{-14} A

Transducer: electron multiplier. Average current gain, $\bar{G} = 10^5$.

Signal conditioner: current-follower amplifier. Feedback resistance, $R = 10^6\ \Omega$ (temperature coefficient = $5 \times 10^{-5}\,°C^{-1}$, rms variation of average temperature = 0.4°C); current noise density, $i_n = 2 \times 10^{-16}\ A/\sqrt{Hz}$[34]; voltage noise density, $v_n = 10^{-6}\ V/\sqrt{Hz}$[34]; bandwidth, $\Delta f = 10^3$ Hz.

Digitizer: analog-to-digital converter. Full scale voltage, $V_{fs} = 10$ V; 12-bit resolution; digitization rate = 2 kHz or more.

Let us calculate the expected signal-to-noise ratio for such a system when the input ion current is 10^{-12} A.

Signal: From equation (2), $S_v = 0.100$ V

Ion beam noise: From equation (7), $\sigma_{Bv}^2 = 3.20 \times 10^{-6}\ V^2$, $\sigma_{Bv} = 1790$ μV.

Transducer noise: From equation (16), $\sigma_{Xvm}^2 = 3.20 \times 10^{-6}\ V^2$, $\sigma_{Xvm} = 1790\ \mu$V.

Signal-conditioning noise: From equation (18), Johnson noise contributed by the feedback resistor, $\sigma_{Cvj}^2 = 1.66 \times 10^{-11}\ V^2$, $\sigma_{Cvj} = 4.1\ \mu$V; from equation (21), noise due to variations in the average temperature of the feedback resistor, $\sigma_{Cvt}^2 = 4.0 \times 10^{-12}$ V^2, $\sigma_{Cvt} = 2.0$ μV; from equation (22), amplifier noise, $\sigma_{Cva}^2 = 1.0 \times 10^{-9}$ V^2, $\sigma_{Cva} = 32$ μV; total signal-conditioning noise, $\sigma_{Cv}^2 = 1.01 \times 10^{-9}$ V^2, $\sigma_{Cv} = 32$ μV.

Digitization noise: From equation (44), digitization noise ($d = 10$ V/2^{12} $= 2.44 \times 10^{-3}$ V), $\sigma_D{}^2 = 4.97 \times 10^{-7}$ V^2, $\sigma_D = 705$ μV.

Total noise: From equation (4), $\sigma_S{}^2 = 6.90 \times 10^{-6}$ V^2, $\sigma_S = 2630$ μV.

Signal-to-noise ratio: From equation (1), $S/N = 38.1$.

The ion beam noise, the transducer noise, and the noise due to variations in the average temperature of the feedback resistor are all dependent on the magnitude of the ion current. A complete summary of the expected system performance, therefore, takes the form of a graph, as shown in Figure 6. Because transducer noise, the principal noise source other than shot noise, varies with $i^{1/2}$, as does shot noise, the line tracing expected system performance parallels the line indicating the unavoidable shot-noise limitation. Only at very low currents do other noise sources contribute significantly, causing the performance to drop off somewhat more sharply in the range below 10^{-13} A.

3. Experimental Observations of the Performance of Ion-Current-Measurement Systems

The approach taken in the preceding sections is classical and straightforward. The individual noise terms presented and discussed have been independently investigated and validated. The practicing mass spectroscopist, however, has an understandable interest in comparing this theory with reality by asking whether the performance of actual measurement systems exceeds or falls short of predicted levels. The development of procedures which examine this question without clouding the issue requires some attention.

3.1. Experimental Approach

Conceptually, the overall signal-to-noise ratio of an ion-current-measurement system could be determined by making repeated observations of a perfectly stable ion beam uncorrupted by any noise other than the unavoidable shot noise. In this case, the observed standard deviation of the population of individual signal (ion current) measurements would give the noise directly, and the computation of the signal-to-noise ratio would be trivial. It would be necessary, however, somehow to prove that the ion beam was free of noise other than shot noise. When the range of things that might sabotage perfection is considered in its full magnificence, it is easy to conclude that some more favorable experimental design should be

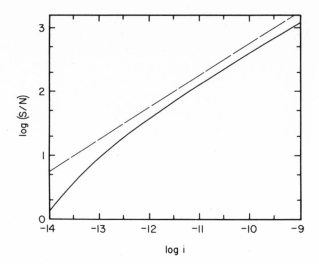

Figure 6. Signal-to-noise ratio as a function of ion current for the system discussed in the text. The broken line indicates the shot-noise limit.

sought, one which would be immune to the effects of noise other than that arising in the measurement system.

Repeated measurements of an ion-current *ratio* can allow satisfactory isolation of noise originating in the ion-current-measurement system. The use of a ratio as the determined quantity has the effect of cancelling most errors due to variations in performance of the ion source and the inlet system. A further increase in purity is obtained by comparing only adjacent ratio observations within a series of measurements, thus minimizing the effect of any long-term drifts which might be influencing background ion currents or other variations not originating in the ion-current-measurement system. The procedure described is similar to that used in measurements of natural variations of stable isotopic abundances,[35,36] and we define the parameter of interest as follows:

$$\delta_i = [(R_i/R_{i+1}) - 1] \times 10^3 \qquad (50)$$

where δ is the relative difference, expressed in parts per thousand, between successive measurements of a ratio R. Following the usage in isotopic analyses, we shall assign the symbol $^o\!/\!_{oo}$, and the term "per mil" to the statistic δ. For any sequence of n measurements of R, it is possible to calculate $n - 1$ values for δ. For such a sequence, the variance in δ is given by

$$\sigma_\delta^2 = \sum_{i=1}^{n-1} \delta_i^2/(n - 1) \qquad (51)$$

where it has been assumed that the true value of δ is known to be zero in this case, since δ represents successive measurements of the same ratio.

The variance in δ is an easily observed quantity, and its value can be compared to that predicted by signal-to-noise ratio calculations in order to evaluate the utility of the approach described in the first part of this review. The variance in δ can be related to the signal-to-noise ratio as follows. We note, first, that the variance in the measurement of a ratio of two signals, $R = S_1/S_2$ is given by

$$\sigma_R{}^2 = \left(\frac{\partial R}{\partial S_1}\right)^2 \sigma_{S_1}{}^2 + \left(\frac{\partial R}{\partial S_2}\right)^2 \sigma_{S_2}{}^2 \tag{52}$$

or, in terms of the relative variances, by

$$(\sigma_R/R)^2 = (\sigma_{S_1}/S_1)^2 + (\sigma_{S_2}/S_2)^2 \tag{53}$$

The variance in δ is given by

$$\sigma_\delta{}^2 = \left(\frac{\partial \delta}{\partial R_i}\right)^2 \sigma_{R_i}{}^2 + \left(\frac{\partial \delta}{\partial R_{i+1}}\right)^2 \sigma_{R_{i+1}}{}^2 \tag{54}$$

Evaluating the derivatives and assuming that $R_i \approx R_{i+1}$ and that $\sigma_{R_i} \approx \sigma_{R_{i+1}}$, we obtain

$$\sigma_\delta{}^2 = 2 \times 10^6(\sigma_R/R)^2 = 2 \times 10^6[(\sigma_{S_1}/S_1)^2 + (\sigma_{S_2}/S_2)^2] \tag{55}$$

in which we find that σ_δ is given in terms of the inverse signal-to-noise ratios.

Because it is based on differential ratio measurements, the approach described here does not allow determination of the signal-to-noise ratios of individual ion currents. For example, given σ_δ, equation (55) cannot be solved for σ_{S_1} and σ_{S_2} without making assumptions about their relative magnitudes. The determination of the values of the various terms contributing to σ_{S_1} and σ_{S_2} is even more remote. Based on the signal-to-noise expressions given in the preceding sections, however, it is possible to calculate expected values of σ_δ and to examine the experimental results for agreement. It is this approach which will be taken here.

3.2. Observed Levels of Performance

3.2.1. Ion Counting

Schoeller and Hayes[30] have shown that the performance of an ion counting system is, in fact, limited only by ion beam shot noise, provided that limitations imposed by deadtime uncertainties[32] are observed. In spite of this favorable result, ion-counting systems are not widely used for high-precision isotope ratio measurements because the analysis time is quite long. Since coincident count losses limit the dynamic range of the system to relatively small ion currents, long observation times are required to collect enough ions for high-precision measurements. For instance, the times required to achieve precisions of 1.0, 0.5, and 0.1‰ with an ion counting mass spectrometer measuring isotopic ion currents of 1.6×10^{-13} and 1.9×10^{-15} A (10^6 and $\sim 10^4$ ions/s, respectively) are 10.5 min, 42 min, and 17 h, respectively.[30]

3.2.2. Faraday Cup, Current Follower

Repeated observations of δ_i for such a system were obtained using the same mass spectrometer employed by Schoeller and Hayes in their ion-counting work,[30] but replacing the electron multiplier with a Faraday cup. The expected signal values and noise contributions can be determined as follows:

S_1(m/e 45 of CO_2), from equation (2) with $i = 1.6 \times 10^{-13}$ A, $\bar{G} = 1.0$, and $R = 10^{11}$ Ω; S_2(m/e 44 of CO_2) calculated in the same way, but with $i = 1.3 \times 10^{-11}$ A.

σ_{Bv}^2, from equation (7)

$\sigma_{Xf}^2 = 0$

$\sigma_{Cv}^2 : \sigma_{Cvj}^2$ from equation (18) with $T = 300°$; σ_{Cvfa}^2 from equation (22) with $i_n = 1.4 \times 10^{-16}$ A/$\sqrt{\text{Hz}}$ and $v_n = 1.4 \times 10^{-5}$ V/$\sqrt{\text{Hz}}$

σ_D^2 from equation (45) with $d = 8.14 \times 10^{-5}$ V for S_1 and $d = 7.81 \times 10^{-4}$ V for S_2, these values being obtained by the use of a voltage to frequency converter ($V_{fs} = 10$ V, $F = 10^6$ Hz) with incremental counting times of 122.88 ms for S_1 and 12.80 ms for S_2.

Each of the expressions cited requires the specification of an integration time t or, in the case of the digitization noise, the specification of n, the total number of counting intervals. These parameters were varied, and σ_δ was determined as a function of N_{45}, the total number of mass 45 ions collected. The signal integration times, t_1 and t_2, are related to N_{45} by the expression

$$t_1 = 9.6t_2 = N_{45}e/i_1 \tag{56}$$

where unequal observation times were chosen so that $t_1/t_2 \sim 1/\sqrt{R}$, as required for minimizing the uncertainty in R within a given total observation time (the origin of this requirement is discussed in detail in the concluding section of this review). The number of counting intervals, n, is given by

$$n = t_1/0.12288 = t_2/0.01280 \tag{57}$$

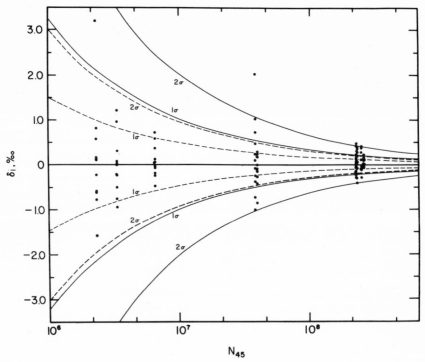

Figure 7. Observed values of δ_i (equation 50) vs. N_{45}, the number of m/e 45 ions collected (see equation 56) at the Faraday cup. The broken contours indicate the scatter expected for shot-noise limiting only. The solid contours have been calculated from system characteristics as indicated in the text.

The results of this series of measurements are summarized in Figure 7, in which the points represent individual measurements of δ and the contours represent expected levels of performance. If ion-beam shot noise alone were responsible for the scatter in the results, 68% of the results should fall within the dashed contours labeled "1σ," and 95% should fall within the dashed contours labeled "2σ." The unbroken contours represent expected degrees of scatter predicted on the basis of the signal-to-noise calculations summarized here [equation (55)]. Experimentally, 84 of 108 observations (77.8%) lie within the 1σ contours predicted on the basis of signal-to-noise calculations, while 105 of 108 (97.2%) lie within the 2σ contours. When the contours based on shot noise alone are considered, the corresponding fractions are 60.6% and 84.8%. It is apparent that the actual performance of the system is more accurately predicted by the signal-to-noise calculations than by ion-statistical calculations alone.

A well-defined comparison between the experimental data and the various predictions can be based upon the F test. When the 53 observations of δ at $N_{45} \sim 2 \times 10^8$ are considered, we find

$$F = \sigma_\delta^2 \,[\text{observed}]/\sigma_\delta^2 \,[\text{predicted by shot noise only}] = (0.199/0.100)^2$$

a value which is significant at the 99% confidence level. When the observed precision is compared to that predicted by signal-to-noise ratio calculations, no significant difference is found.

3.2.3. Electron Multiplier, Current Follower

Parallel tests were made using an electron multiplier as the transducer in the same mass spectrometer system, the only additional modification being the use of a different current-follower amplifier with a lower feedback resistance. Calculations of expected performance levels duplicate those outlined for the Faraday cup system, with only the following parameters being changed: $\bar{G} = 10^6$, $R = 10^5$ Ω, and $i_n = 1.4 \times 10^{-14}$ A/$\sqrt{\text{Hz}}$; and with σ_{Xvm}^2 being calculated from equation (16).

The results of this series of measurements are summarized in Figure 8 together with performance contours calculated as for Figure 7. Of a total of 88 observations, respectively, 76.1% and 97.7% fall within the 1σ and 2σ contours predicted on the basis of calculated signal to noise ratios. When the F test is employed, the deviation from precision expected on the basis of ion-beam shot noise alone is significant at the 99% confidence level, but no statistically significant difference is found when the observed precision is compared to that predicted from signal-to-noise ratio calculations.

3.2.4. Faraday Cup, Capacitive Integration

Beckinsale et al.[37] have reported the results of performance tests of the system developed by Jackson and Young.[26] It was observed that the precision obtained in actual measurements was slightly worse than that calculated on the basis of the S/N model or of ion beam shot noise alone. While the differences in precision are statistically significant, it should be emphasized that this system does achieve a higher level of performance than the analog systems previously described, and that no attempt was made in the calculations of expected performance to allow for the effects of temperature variations.

4. Discussion

A few very general comments and conclusions can be based upon the information summarized in the preceding sections of this review. We will

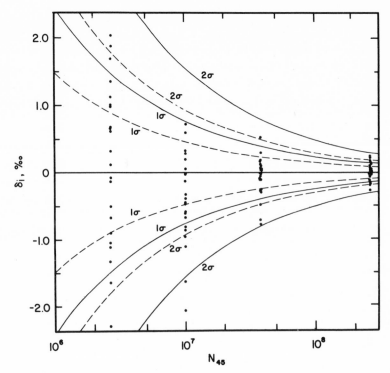

Figure 8. Observed values of δ_i (equation 50) vs. N_{45}, the number of m/e 45 ions collected (see equation 56) at an electron multiplier transducer. The broken contours indicate the scatter expected for shot-noise limiting only. The solid contours have been calculated from system characteristics as indicated in the text.

be brief, and, except where specifically indicated, will concern ourselves entirely with current-follower systems, excluding capacitive integration systems from further discussion.

4.1. General Features of Performance Curves

The relationship between ion current and the signal-to-noise ratio for any measurement system can be placed in the form

$$S/N = (S^2/N^2)^{1/2} = [ki^2/(ai^2 + bi + c)]^{1/2} \tag{58}$$

where k, a, b, and c are constants related to the specifications of the system and where the noise terms in the denominator have been collected in three categories according to their dependence on ion current. Among the noise sources considered here, only the effects of temperature variations (equation 20) vary directly with i ($\sigma^2 \propto i^2$). Both shot noise [equation (7)] and electron multiplier secondary electron emission noise [equation (16)] are proportional to $\sqrt{i}$ ($\sigma^2 \propto i$). Finally, Johnson noise [equation (18)], digitization noise [equation (45)], and amplifier noise [equation (22)] all make constant noise contributions, independent of the ion current level. Clearly, the relative magnitudes of the various terms in the denominator can vary significantly within the range of ion currents to which any system is applied.

The generalized effects of the various types of noise are graphically summarized in Figure 9. The dominant feature of this (or of any specific performance curve) is the limitation imposed by ion-current shot noise. The $1/\sqrt{i}$ dependence gives rise to a line with a slope of 1/2 on a graph of log (S/N) vs. log i. When an electron multiplier is used as the transducer in a system, the effects of secondary emission noise will reduce the performance by a further constant factor, giving a line parallel to and below the shot-noise limit on a graph of this type. Shot noise and secondary emission noise are the only important noise terms for many practical electron multiplier systems, and the performance curves for these systems (e.g., Figure 6) are relatively featureless. The fixed-magnitude noise sources, particularly Johnson noise, are more likely to be of importance in Faraday cup systems, although such systems are free of secondary emission noise and, in their favorable ranges of application, can approach the shot-noise limit more closely than electron multiplier systems (contrast the heavy solid and dashed lines in Figure 10, to be discussed in detail below). Where constant noise sources are dominant, the slope of a log (S/N) vs. log i curve is 1.0. The final category of noise terms, those which are directly propor-

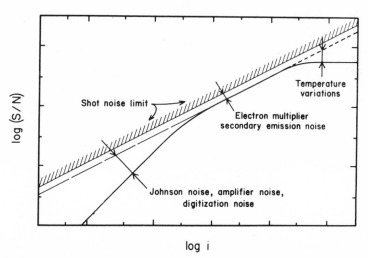

Figure 9. Generalized system performance curve indicating the effect of various noise sources.

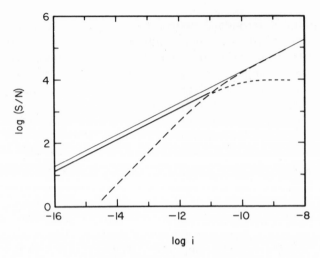

Figure 10. Performance curves for typical ion-current-measurement systems incorporating as the transducer an electron multiplier (heavy solid line) or a Faraday cup (heavy broken line). The thin continuous line indicates the shot noise limit. The systems are compared at a common bandwidth of $\Delta f = 1$ Hz, and the possible additional limitations due to digitization noise have been excluded. Specifications for the electron multiplier system: $R = 10^7$ Ω, $\bar{G} = 3 \times 10^5$, $i_n = 2 \times 10^{-16}$ A/$\sqrt{\text{Hz}}$, and $v_n = 10^{-6}$ V/$\sqrt{\text{Hz}}$. Specifications for the Faraday cup system: $R = 10^{10}$ Ω, $i_n = 10^{-15}$ A/$\sqrt{\text{Hz}}$, $v_n = 10^{-5}$ V/$\sqrt{\text{Hz}}$. The heavy dotted line shows the effect of temperature variations on the Faraday cup system assuming $\alpha = 1000$ ppm/C° and $\sigma_T = 0.1$ C°.

tional to signal level, is most devastating in its effect. As noted in Figure 9, the effects of temperature variations can firmly limit the signal-to-noise ratio, causing the performance curve to flatten completely. Although temperature variations can be minimized, if not completely controlled, in the laboratory environment, this is not always convenient in the field or, especially, in spacecraft. In any case, it must be kept in mind that a typical temperature coefficient for a 10^{10}-Ω resistor is the order of 1000 ppm/C°,[25] and that the required precision of temperature control is, accordingly, very great.

While we have demonstrated that temperature variations of the feedback resistance can severely limit the S/N of an ion-current measurement, there are many measurement situations where temperature variations are relatively unimportant. Experiments based on relative ion-current abundances are not subject to severe temperature-induced noise if the measurement time for the individual ion currents is short compared to the thermal time constant of the resistor. Measurements requiring the use of multiple collector systems can minimize temperature-induced variations by employing feedback resistances with matched temperature coefficients.

4.2. Comparisons between Systems

It is impossible to attribute overall superiority to any particular system. It should be evident that the characteristics of a system are determined not only by the transducer employed, but also, in very large part, by the other components in a system. Thus, even for a specific application, one system employing an electron multiplier might offer better performance than another employing a Faraday cup, while that Faraday cup system might, in turn, substantially outperform another which happened also to use an electron multiplier as the transducer. Having offered this warning, we will make a few general comparisons and comments.

4.2.1. Attainable Precision

Two specific and typical systems, one employing an electron multiplier and the other a Farady cup, are compared in Figure 10 (details and specifications are given in the caption). The performance curves are terminated at an output signal level of 30 V. It happens that this corresponds to an input ion current of 10^{-11} A for the electron multiplier system and that, at this point, the two systems offer equal signal-to-noise ratios. For ion currents greater than 10^{-11} A, the particular Faraday cup system specified here offers better performance than could any electron multiplier system. The relatively higher performance of the Faraday cup system could be extended to lower input ion currents through the use of a larger feedback resistor in the current-follower amplifier, but this is usually avoided because the time constant of the system becomes inconveniently long. By the same token, the use of the electron multiplier could be extended to ion currents substantially in excess of 10^{-11} A through the use of a lower feedback resistance or by lowering the current gain of the multiplier. Often, however, this is avoided because the simpler Faraday cup system offers equal or better performance and complete freedom from systematic gain variations. It happens, therefore, that 10^{-11} A forms a boundary—to the extent that any such boundary can be suggested—between the typical ranges of application of these transducers. When the boundary is ignored, it is generally because other conditions, usually relating to the response time or stability of the system, require it.

4.2.2. Bandwidths

The bandwidths of the ion-current-measurement systems described cover a very wide range, with ion-counting systems being capable of responding in less than a microsecond and capacitive integration systems usually employing integration times on the order of 10 s or more. For the current-follower systems, the minimum response time is usually set by the minimum RC time constant attainable. For electron multiplier systems this can usually be held to any desired value, furnishing bandwidths up to at least 10^4 Hz. For Faraday cup systems, on the other hand, R is seldom less than 10^9 Ω and the minimum value of C is effectively set by the stray capacitance in the collector system, which is seldom less than 10 pF. Thus, Faraday cup systems with $R = 10^{10}$ Ω are ordinarily restricted to applications in which $\Delta f \leq 1$ Hz can be tolerated.

4.2.3. Selection of an Ion-Current-Measurement System

The selection of an optimal system for a specific application requires thorough consideration of the range of ion currents to be measured, the required speed of response, and the precision and accuracy desired. When these requirements have been defined and prioritized, the selection of the "best" system is straightforward, if not easy.

Applications requiring rapid response ($\Delta f > 1$ Hz) generally dictate the use of an electron multiplier. The range of currents to be measured will then determine whether ion counting or analog measurement is appropriate. Our experience indicates that ion counting is definitely to be preferred in the range below 10^{-14} A, that the two alternatives offer approximately equal performance in the range between 10^{-14} and 2×10^{-13} A, and that analog measurement should probably be preferred in the range above 2×10^{-13} A. Ion counting is preferred in the lowest range because of its superior signal-to-noise ratio and freedom from drifts and other systematic errors. Ion counting is not favored in the highest range because errors in deadtime determination can lead to serious inaccuracies, because the performance of some systems is distinctly nonlinear, and because the use of the relatively high gain required can seriously shorten multiplier lifetime at these high current levels.

Where rapid response is not required, any of the transducers mentioned can be employed. In practice, this means only that Faraday cup systems can be suggested for use in the range above 10^{-11} A, with the other preferences indicated in the preceding paragraph being unchanged. Because all the boundaries mentioned are rather diffuse, it might be well to emphasize this fact by noting that an effective measurement system might combine an ion-counting electron multiplier for use in the range below 3×10^{-13} A and a Faraday cup ($R \sim 10^{11}$ Ω) for use in the range above 3×10^{-13} A. While such a system could be outperformed in some respects in the intermediate range around 3×10^{-13} A, its relative freedom from systematic gain variations might be attractive in some applications.

4.2.4. Selection of a Digitization Device

As noted in the section describing digitization noise, it is important that the noise amplitude in the final analog signal be equal to or greater than $d/2$, where d is the quantization interval of the digitization device. If this condition is not met, the signal tends to stay entirely within a single digitization interval, the digital output remains constant or nearly so, effective signal averaging cannot occur [that is, equations (44) and (45) will be invalid], and systematic errors will be dominant.

To consider how it can be assured that σ_S will equal or exceed $d/2$, we can note, first, that σ_S must always equal or exceed σ_B, the ion-beam shot noise. For current-follower systems, then, we can write

$$\sigma_S{}^2 \geq \sigma_{Bv}^2 = 2ei(R\bar{G})^2\Delta f \qquad (59)$$

If we now require that $d/2 \leq \sigma_{Bv}$, we can write

$$d^2 \leq 8ei(R\bar{G})^2\Delta f' \qquad (60)$$

where, in this case, $\Delta f'$ represents the bandwidth of the analog circuitry, which may be equal to or greater than the overall measurement bandwidth. Second, we note that, for any given system with fixed values of R, $\bar{G}$, and d, the requirement for effective signal averaging will be met provided that

$$i \geq d^2/8e(R\bar{G})^2\Delta f' \qquad (61)$$

It is convenient to summarize this situation graphically. Figure 11 plots i as a function of $\Delta f'$ for three different systems in which it has been assumed that $R\bar{G} = 10^{11}\,\Omega$ and in which different values of d, corresponding to the use of various digitization devices, have been assigned.

When an integrating digitizer (for example, a device combining a voltage-to-frequency converter and a counter) is employed, the value of $\Delta f'$ is controlled by the integration time of the digitizer, with $\Delta f' = 1/2\Delta t$, where Δt is the digitization time interval [in this statement we assume $\Delta t \geq 1/2$(bandwidth of the analog circuitry preceding the digitizer)]. Equation (43) then takes the form

$$d = V_{fs}/F\Delta t = 2V_{fs}\Delta f'/F \qquad (62)$$

where V_{fs} is the full-scale input voltage of the voltage-to-frequency converter and F is the corresponding output frequency. Substituting in equation (61), we obtain

$$i \geq [(V_{fs}/FR\bar{G})^2(1/2e)]\Delta f' \qquad (63)$$

a relationship in which i is directly proportional to $\Delta f'$, and which, therefore, contrasts quite strongly with equation (61). If typical values (see figure

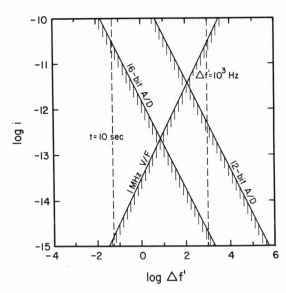

Figure 11. Minimum ion current required for effective signal averaging as a function of $\Delta f'$, the bandwidth of the analog circuitry. The lines have been calculated from equations (61) and (63) assuming that $R\bar{G} = 10^{11}$ Ω and that the full-scale input voltage for the digitizers is 10 V. The line designated "1 MHz V/F" represents an integrating digitization system comprised of a voltage-to-frequency converter (1 MHz full-scale output frequency) and a counter. The lines designated "16-bit A/D" and "12-bit A/D" refer to nonintegrating analog to digital converters with fixed quantization intervals of 153 μV and 2.44 mV, respectively.

caption) are inserted in equation (63), the line designated "1 MHz V/F" in Figure 11 is obtained. It is, for example, apparent that a 1-MHz V/F converter system can be effectively employed to digitize any ion currents greater than 1.6×10^{-15} A when the counting time for each digital output is 10 s.

When a nonintegrating digitizer is employed, the value of $\Delta f'$ is controlled entirely by the analog circuitry and by any time constants associated with the collector itself. In this case, consideration of the magnitude of $\Delta f'$ is especially important in selecting a digitization device. For example, almost any Faraday cup system is likely to have log $\Delta f' \leq 0$ and Figure 11 shows that, in these circumstances, the use of a 12-bit A/D is likely to introduce serious systematic errors. When $\Delta f' = 1$ Hz, even a 16-bit A/D will not offer effective signal averaging for ion currents below 1.8×10^{-12} A. The analog-to-digital converter systems *can* be effectively employed [i.e., equation (45) will be valid] whenever the i, $\Delta f'$ point describing the system lies above the line indicating the critical ion current. For example, if the bandwidth of the analog circuitry is 1 kHz, a 12-bit A/D converter can be effectively employed for the digitization of ion currents greater than 4.5×10^{-13} A, while a 16-bit A/D converter can be used effectively down to ion current levels of 1.8×10^{-15} A.

4.3. Signal-to-Noise Ratios and Optimal Observation Times

Consider the measurement of a ratio $R = S_1/S_2$. Equation (53) relates the relative uncertainty in R to the signal-to-noise ratios for S_1 and S_2. If we let $P = \rho_R/R$ and define X and Y as the signal-to-noise ratios for S_1 and S_2 when the integration times are each 1 s, that is

$$X \equiv (S_1/\sigma_{S_1})_{t=1} \quad \text{and} \quad Y \equiv (S_2/\sigma_{S_2})_{t=1}$$

we can write

$$(S_1/\sigma_{S_1})_{t=t_1} = \sqrt{t_1}X \quad \text{and} \quad (S_2/\sigma_{S_2})_{t=t_2} = \sqrt{t_2}Y$$

to represent the signal-to-noise ratios at any arbitrary integration times t_1 and t_2. [Note that we *assume* $(S/N) \propto \sqrt{t}$, thus requiring that σ_{Cvt} can be neglected and that $\sigma_D{}^2 \propto 1/t$, conditions which might not always apply.] Then, for any total measurement time, $t = t_1 + t_2$, we can write

$$P^2 = 1/t_1X^2 + 1/t_2Y^2 \tag{64}$$

Letting r represent the ratio of observation times, that is, $t_1 = rt_2$, we write

$$P^2 = 1/rt_2X^2 + 1/t_2Y^2 \tag{65}$$

Solving for t_2, we obtain

$$t_2 = 1/rP^2X^2 + 1/P^2Y^2 \tag{66}$$

an expression for the total analysis time is then given by

$$t = t_1 + t_2 = 1/P^2X^2 + r/P^2Y^2 + 1/rP^2X^2 + 1/P^2Y^2 \tag{67}$$

setting $\partial t/\partial r = 0$ in order to minimize t we obtain

$$r = t_1/t_2 = Y/X = (S/N)_2/(S/N)_1 \tag{68}$$

thus finding that the relative observation times should be inversely proportional to the signal-to-noise ratios. For a shot-noise-limited measurement we have $(S/N) \propto 1/\sqrt{i}$ and thus find $r = 1/\sqrt{R}$, in accordance with the familiar rule that, for counting measurements, the observation times should be inversely proportional to the square roots of the count rates. Equation (68) is more broadly applicable, allowing the determination of the optimal ratio of observation times for any pair of signals for which $(S/N) \propto \sqrt{t}$, without requiring that the measurement be limited only by shot noise.

ACKNOWLEDGMENTS

The first careful considerations of signal-to-noise ratios in our laboratories were summarized in unpublished sections of the thesis of Dr. Dale A. Schoeller, now a member of the faculty of the Pritzker School of Medicine at the University of Chicago. The beginning sections of this review represent evolutions of Prof. Schoeller's work, and we are happy to acknowledge his contributions. We appreciate the assistance of E.A. Vogler and B. Enos in deriving equation (41) and the comments of Drs. Robert Beckinsale and Leonard A. Dietz, and we are grateful for many helpful discussions with our colleague and editor, Prof. G. M. Hieftje. The development of the isotope ratio mass spectrometry laboratories at Indiana University has been funded by the National Institutes of Health (GM-18979) and the National Aeronautics and Space Administration (NGR 15-003-118), whose support is greatly appreciated.

5. References

1. E. B. Wilson, Jr., *An Introduction to Scientific Research*, pp. 191–195, McGraw-Hill, New York (1952).
2. C. Brunnée, Concerning ion reflection and secondary electron emission due to the impact of alkali ions on pure Mo surfaces, *Z. Phys.*, *147*, 161–183 (1957).
3. E. R. Cawthron, D. L. Cotterell, and M. Oliphant, The interaction of atomic particles with solid surfaces at intermediate energies. II Scattering processes, *Proc. R. Soc. London Ser. A. 314*, 53–72 (1968).
4. J. S. Allen, Detection of single positive ions, electrons and photons by a secondary electron multiplier, *Phys. Rev., 55*, 966–971 (1939).
5. J. S. Allen, An improved electron multiplier particle counter, *Rev. Sci. Instrum., 18*, 739–749 (1947).
6. J. S. Allen, Preliminary Report No. 10, Nuclear Science Series, National Research Council (1950).
7. G. W. Goodrich and W. C. Wiley, Resistance strip magnetic electron multiplier, *Rev. Sci. Instrum., 32*, 846–849 (1951).
8. G. W. Goodrich and W. C. Wiley, Continuous channel electron multiplier, *Rev. Sci. Instrum., 33*, 761–762 (1962).
9. H. Becker, E. Dietz, and U. Gerhardt, Preparation and characteristics of a channel electron multiplier, *Rev. Sci. Instrum., 43*, 1587–1589 (1972).
10. Specification, ITT Type F4074, Electro-Optical Produces Division, ITT, Fort Wayne, Indiana.
11. C. La Lau, Mass discrimination caused by electron multiplier detector, in: *Topics in Organic Mass Spectrometry* (A. L. Burlingame, ed.), pp. 93–121, Wiley-Interscience, New York, (1970).
12. M. Van Gorkom and R. E. Glick, Electron multiplier response under positive ion impact, I. Secondary electron emission coefficients, *Int. J. Mass Spectrom. Ion Phys., 4*, 203–218 (1970).
13. M. Van Gorkom, D. P. Beggs, and R. E. Glick, Electron multiplier response under

positive ion impact, II. Secondary pulse height distributions, *Int. J. Mass Spectrom. Ion Phys.*, *4*, 441–450 (1970).

14. R. C. Lao, R. Sander, and R. F. Pottie, Discrimination in electron multipliers for atomic ions, I. Multiplier yields for 24 mass-analyzed ions, *Int. J. Mass Spectrom. Ion Phys.*, *10*, 309–313 (1972).

15. R. F. Pottie, D. L. Cocke, and K. A. Gingerich, Discrimination in electron multipliers for atomic ions. II. Comparison of yields for 61 atoms, *Int. J. Mass Spectrom. Ion Phys.*, *11*, 41–48 (1973).

16. R. W. Engstrom and E. Fischer, Effects of voltage divider characteristics on multiplier phototube response, *Rev. Sci. Instrum.*, *28*, 525–527 (1957).

17. L. A. Deitz, Basic properties of electron multiplier ion detection and pulse counting methods in mass spectrometry, *Rev. Sci. Instrum.*, *36*, 1763–1770 (1965).

18. L. A. Dietz, L. R. Hanrahan, and A. B. Hance, Single electron response of a porous KCL transmission dynode and application of Polya statistics to particle counting in an electron multiplier, *Rev. Sci. Instrum.*, *38*, 176–183 (1967).

19. L. A. Dietz, General method for computing the statistics of charge amplification in particle and photon detectors used for pulse counting, *Int. J. Mass Spectrom. Ion Phys.*, *5*, 11–19 (1970).

19a. L. A. Dietz and J. C. Sheffield, Secondary electron emission induced by 5-30-KeV monatomic ions striking thin oxide films, *J. Appl. Phys.*, *46*, 4361–4370 (1975).

20. D. A. Schoeller, PhD thesis, A Computer Controlled Ion Counting Isotope Ratio Mass Spectrometer, Department of Chemistry, Indiana University (1974).

21. J. R. Prescott, A statistical model for photomultiplier single-electron statistics, *Nucl. Instrum. Methods*, *39*, 173–179 (1966).

22. D. M. Hunten, *Introduction to Electronics*, Holt, Rinehart, and Winston, New York (1964).

23. J. D. Ingle, Jr., and S. R. Crouch, Critical comparison of photon counting and direct current measurement techniques for quantitative spectrometric methods, *Anal. Chem.*, *44*, 785–794 (1972).

24. L. F. Herzog and T. J. Eskew, Voltage coefficients of high value resistors, Proc. 23rd Ann. Conf. on Mass Spectrometry and Allied Topics, pp. 262–265, American Society for Mass Spectroscopy, (1975).

25. (a) Specifications, Type RX-1 Hi-Meg Resistors Victoreen Instrument Division, Cleveland, Ohio. (b) Specification, Type M51, Dale Electronics, Inc., Columbus, Nebraska.

26. M. C. Jackson and W. A. P. Young, A capacitive integration system for the precise measurement of isotopic ratios in a mass spectrometer, *Rev. Sci. Instrum.*, *44*, 32–34 (1972).

27. A. J. Diefenderfer, *Principles of Electronic Instrumentation*, W. B. Saunders, Philadelphia (1972).

28. P. C. Kelly and G. Horlick, Practical considerations for digitizing analog signals, *Anal. Chem.*, *45*, 518–526 (1973).

29. N. R. Daly, Scintillation type mass spectrometer ion detector, *Rev. Sci. Instrum.*, *31*, 264–267 (1960).

30. D. A. Schoeller and J. M. Hayes, Computer controlled ion counting isotope ratio mass spectrometer, *Anal. Chem.*, *47*, 408–414 (1975).

31. J. M. Hayes, D. E. Matthews, and D. A. Schoeller, Effective deadtime of pulse counting detector systems, *Anal. Chem.*, *50*, 25–32 (1978).

32. J. M. Hayes and D. A. Schoeller, High precision pulse counting: Limitations and optimal conditions, *Anal. Chem.*, *49*, 306–311 (1977).

33. S. D. Conte and C. de Boer, *Elementary Numerical Analysis*, p. 33, 2nd ed., McGraw-Hill, New York, (1972).

34. Specifications, Analog Devices Type 310 Varactor Bridge Electrometer, Analog Devices, Norwood, Massachusetts.

35. H. Craig, The geochemistry of the stable carbon isotopes, *Geochim. Cosmochim. Acta,* *3,* 53–91 (1953).

36. C. R. McKinney, J. M. McCrea, S. Epstein, H. A. Allen, and H. C. Urey, Improvements in mass spectrometers for the measurement of small differences in isotope abundance ratios, *Rev. Sci. Instrum., 21,* 724–730 (1950).

37. R. D. Beckinsale, N. J. Freeman, M. C. Jackson, R. E. Powell, and W. A. P. Young, A 30 cm radius double collecting mass spectrometer with a capacitor integrating detector for high precision isotopic analysis of carbon dioxide, *Int. J. Mass Spectrom. Ion Phys., 12,* 299–308 (1973).

6

Analytical Techniques for the Study of Biological Membranes

Kuang-Pang Li and Feng-Shyong S. Chuang

1. Introduction

Molecular studies of membrane biology have received extremely enthusiastic attention in recent years. This is in part due to the increasing awareness of membrane involvement in many aspects of life phenomena, such as neural transport, muscle action, metabolism, carcinogenesis, immunology, etc., and in part due to the rapid development of highly delicate analytical techniques for studying macromolecular phenomena. With these techniques much information not available a decade ago is now readily accessible. Although a thorough understanding of membrane biology is still far from reached, the information acquired so far does discern, more or less, the gross aspects of the mysterious puzzle.

Several books[1-5] and many reviews[6-13] have appeared recently covering various aspects and applications of membrane science. Unfortunately, most of them and numerous research papers were written mainly for biological and medical scientists. To people unfamiliar these fields, these publications are, sometimes, more confusing than educational. It is, therefore, not the intention of this chapter to review past work in an exhaustive manner, nor to compile a large body of facts, figures, and references. Instead, we would like to stress the various analytical methods most useful for membrane studies. Basic principles of these methods have been discussed extensively in a variety of textbooks and review articles; emphasis

Kuang-Pang Li • Department of Chemistry, University of Florida, Gainesville, Florida 32611 Feng-Shyong S. Chuang • Physical and Analytical Section, Norwich Pharmaceutical Co., Norwich, New York 13815

253

Kuang-Pang Li and Feng-Shyong S. Chuang

is, therefore, placed mostly upon modifications which are specific for membrane researches.

The chapter can be divided into three parts. The gross structure of biomembranes is briefly reviewed in terms of a universally accepted model, the fluid mosaic model, to provide essential background knowledge. This background is then followed by discussions of various analytical techniques used for the study of membrane organization and membrane transport. Other important biofunctions such as the membrane involvement in cell–cell interactions, cell differentiation, growth control, and immunology are too biological for most chemists and are beyond the scope of this chapter. Methods used for the study of these latter functions are therefore intentionally omitted.

2. Structure of Biomembranes—The Fluid Mosaic Model

Due to the high degree of specialization in membrane biofunctions, there is also a great diversity of membrane organization. Each kind of membrane has its own special composition and unique structural arrangement, and there do not seem to be any useful generalizations to be made even about gross membrane structure. However, the overall arrangement of the principal constituents in an intact membrane can be generalized. Such generalizations, or idealized model, are very useful in understanding the characteristics and functions of membranes. They may also be useful to explain old experiments and suggest new ones.

Several hypothetical membrane models have been proposed.[14-22] Of these models, the fluid mosaic model developed by Singer and Nicolson[23] seems to be the most satisfactory. It brings together into a single theory a new dynamic concept of the organization of cell membranes. The following discussion follows the same line of reasoning as in Singer's development.

Most biological cells and cell organelles live in an aqueous environment. From the energetic point of view, the functional membranes of these cells and organelles, including both plasmalemmal and intercellular membranes, must be in a thermodynamically most favorable state in such an environment. This implies that the molecular arrangement of membranes must be one of the lowest free energy state, determined mainly by the membrane and the aqueous environment.

There are three major classes of components in biological membranes, namely, proteins, lipids (mainly phospholipids), and oligosaccharides. The relative proportion of these components varies widely according to species and membrane classes. Generally, the proteins are predominant and play an important role in determining the membrane structure. All these compounds are amphiphatic, containing in the same molecule both nonpolar and polar regions. In order to attain the lowest free energy state in an aqueous environment, their hydrophobic nonpolar regions, i.e., the nonpolar acid residues of the proteins and the fatty acid chains of the phospholipids, should be sequestered to the maximum extent feasible from contact with water, while their hydrophilic polar regions, i.e., the ionic and polar groups of the proteins, the lipids and the oligosaccharides, should be in direct contact with the aqueous phase. Other noncovalent interactions such as hydrogen bonding, dipole–dipole, and electrostatic interactions should also contribute to the stabilization of the membrane structure. However, from energy considerations, they are very likely of secondary magnitude compared to the overwhelming hydrophilic and hydrophobic interactions.

Studies[24,25] on phospholipid bilayers, which are found in a variety of intact membranes, have confirmed the above predictions. In these structures the fatty acid chains are sequestered together away from contact with water and the ionic and zwitterionic groups are in direct contact with the aqueous phases at the exterior surfaces of the bilayer. The thickness of such a phospholipid bilayer with its fatty chains in a noncrystalline state is about 40–45 Å.[26] The average total thickness of membranes is about 75 Å as taken from the density distribution in $KMnO_4$-fixed specimens.[15] This leaves only 15–18 Å space on each side of the bilayer for the distribution of proteins and associated oligosaccharides if they were spread out over the bilayer surfaces.

According to their association with membranes, the membrane proteins can be classified into two general categories, the peripheral and integral proteins. The peripheral, or extrinsic, proteins are those which can be relatively easily dissociated, molecularly intact, from membranes by mild treatments. In their dissociated state they are relatively soluble in neutral aqueous buffer. This implies that they are held to the membrane only by rather weak noncovalent (perhaps mainly electrostatic) bonding and are not strongly associated with membrane lipid. They are of lesser importance and may not be directly relevant to the central problems of membrane structure. Examples of peripheral proteins are mitochondrial cytochrome c and erythrocyte spectrin.[27]

If these proteins were spread over the membrane surface, the size of the membrane allows only a single molecular layer of proteins to exist (the thickness of a single monolayer of polypeptide is about 10 Å).[10] Such a monolayer is evidently not the thermodynamically most favorable arrangement of proteins because the ionic heads of the phospholipid molecules would then be blanketed from contact with the aqueous phase while a significant fraction of the nonpolar residues of the protein molecules would have to be exposed to the water. To maximize the hydrophilic and hydrophobic interactions in membranes, and thus minimize the free energy, the protein molecules are more likely to be globular.[28]

The above argument applies to the integral proteins as well. The latter are membrane proteins which require much more drastic treatment, with reagents such as detergents, bile acids, or protein denaturants, in order to be dissociated from membranes. Usually, lipids remain associated with these solubilized proteins, and if the lipids are extracted, the proteins are then highly insoluble or tend aggregate in aqueous buffer around neutral pH. If these globular proteins were attached to the surface of a lipid bilayer, it would require a membrane thickness much larger than the 75–90 Å generally observed.

Based on these thermodynamic considerations, together with much experimental evidence, Singer and co-workers proposed the hypothetical "fluid mosiac" model of membranes. A pictorial representation of this is shown in Figure 1 and comprises the following features.

1. Biomembranes have a long-range mosaic structure with lipids con-

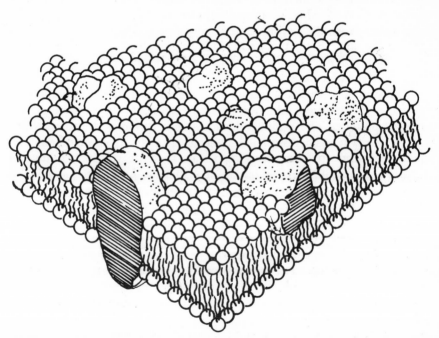

Figure 1. Three-dimensional and cross-sectional views of the fluid mosaic model. The solid bodies with stippled surfaces represent the globular integral proteins, which at long range are randomly distributed in the plane of the membrane. The small circles represent the ionic and polar heads of the phospholipids.

stituting the matrix. The matrix is predominantly arranged as an interrupted bilayer with ionic and polar head groups in contact with the aqueous phase. Under physiological conditions, the bilayers are generally in a fluid rather than a crystalline state (myelin is an exception). In other words, the membrane structure is actually a dynamic rather than a static one.

2. The globular protein molecules (mainly the integral ones), in the form of glycoproteins or lipoproteins, intercalate within the lipid bilayer. The globular proteins are amphiphatic as are the phospholipids. In other words, they are structurally asymmetric with highly polar and nonpolar ends. The polar ends are in contact with water, and the nonpolar ends are embedded in the hydrophobic interior of the membrane.

3. An integral protein molecule with the appropriate size and structure, or a suitable aggregate of integral proteins, may span the entire membrane; that is, it may have regions in contact with the aqueous solvent on both sides of the membrane.

4. The distribution of aggregates of proteins over the entire surface of the membrane is normally random; that is, there is generally no long-range order in the structure. Moreover, proteins are expected to undergo translational diffusion within the membrane, at rates determined in part by the effective viscosity of the lipid, unless they have been tied down by some specific interactions intrinsic or extrinsic to the membrane.

This model provides novel ways of thinking about membrane functions such as nerve impulse transmission, cross-membrane transport, effects of specific drugs on membranes, complex processes of oxidative phosphorylation, photosynthesis, etc., and suggests mechanisms that are amenable to experimental tests. The model may also be employed for the explanation of the mechanism of various membrane-mediated cellular phenomena.

3. Analytical Methods for Membrane Structural Elucidation

In accordance with the fluid mosaic model, biomembranes can generally be regarded as dynamic systems rather than as rigid molecular associations. They are best thought of as two-dimensional viscous solutions. Long-range order is usually not expected to be found under physiological conditions. Because of this dynamic nature, structural elucidation of a membrane is concerned more with the evaluation of the structural variation with time than determination of the actual configuration at a particular moment. In other words, we are more interested in the molecular motion within and the fluidity of a membrane than with the topographical picture of membranes. However, the latter information can be obtained with techniques such as electron microscopy, X-ray diffraction, or freeze-etching.[29]

The time scale of most molecular motions of biological importance in membranes falls in the range of accessibility to NMR, ESR, and fluorescence spectroscopy. These methods are, therefore, ideally suited for such studies. Of these methods, the NMR technique is the least sensitive and requires the largest sample size. However, it has the advantage of employing the naturally occurring membranes as the source of signal. ESR and fluorescence probe techniques, on the other hand, have greater sensitivity and employ simpler instrumentation but require that a nonbiological probe molecule be introduced into the membrane.

3.1. Nuclear Magnetic Resonance

3.1.1. Basic Principles

Most NMR studies of membrane fluidity are basically relaxation measurements.

For simplicity, let's consider a pair of spins of $\frac{1}{2}$, A and B, in the system. A and B may be either of the same kind, such as a pair of protons, or of different kinds, such as a proton and a ^{13}C nucleus. They may be in the same molecule or in different molecules. Because of their intrinsic magnetic moments, these spins precess around the applied magnetic field, H_0, with characteristic Larmor frequencies. The latter depend only on the local magnetic field experienced by the spins and on their nuclear magnetic moments. Now, if the nuclei are allowed to move with respect to each other, the relative motion of one nucleus, say B, will exert an induced magnetic field on A, thereby causing the local magnetic field at A to fluctuate, and vice versa. The frequency of the oscillatory magnetic field at A, when it overlaps the precession frequency of A, can induce a nonradiative transition of A from the upper spin state to the lower state. This kind of nonradiative transition is called *spin-lattice relaxation*. Of course, interactions arising from oscillatory and static magnetic fields produced by B or surrounding magnetic nuclei can also relax spin A. However, the relaxation due to the field fluctuation by molecular motion is more important in membrane studies. Because of spin-lattice relaxation, nuclear spin states have a finite lifetime, T_1, which is the average time for the magnetization along the applied field H_0 to decay to $1/e$ of the initial value. T_1 is sometimes called the *longitudinal relaxation time* because it defines the relaxation of the component of magnetization along the z direction, which is commonly defined as the direction of H_0.

Another commonly used but less well-defined relaxation time is the so-called *spin-spin relaxation time*, sometimes termed the *transverse relaxation time*, T_2. It stems from several processes and is commonly measured in terms of the width of the absorption line at half height, $\Delta\nu_{1/2}$. Readers are encouraged to consult, for example, Abragam's[30] or Slichter's[31] books for a more complete discussion of spin–spin relaxation processes.

Since the relative motion of the nuclei are thermal in origin, they occur over a wide range of frequencies. Corresponding to this range of motions, there will also be a wide range of frequencies in the magnetic field fluctuations. Let us assume that these fluctuations are random, so that if we are interested in some quantity, $f(t)$, fluctuating with time, it will fluctuate about a mean value of zero. To measure how rapidly $f(t)$ fluctuates, we need to know the correlation between the fluctuating quantity $f(t)$ first measured at time t and the same quantity determined at a time τ later, i.e., the correlation between $f(t)$ and $f(t + \tau)$. This relationship can be established by taking the average of the product, $f(t) * f(t + \tau)$ over all time,[30] i.e.,

$$G(\tau) = \overline{f(t) * f(t + \tau)} \tag{1}$$

where the asterisk indicates a complex conjugate, since $f(t)$ is generally a complex function. $G(\tau)$ is called the *autocorrelation function* of $f(t)$ for a stationary random process. It is a function only of τ, the time between the two measurements, and is independent of the time at which we performed our first observation, providing the quantity $f(t)$ is stationary (as it is in biological membranes). When τ is very small, that is, when the first and second measurements are made very close in time, $G(\tau)$ is very close to $\overline{f(t)}^2$. As τ increases, $G(\tau)$ decreases and approaches zero as its limit.

A random function like $f(t)$ can be broken down into a continuum of simple harmonic oscillations and the *spectral density*, $J(\omega)$, of any component with angular frequency ω $(= 2\pi f)$ is simply the Fourier transform of $G(\tau)$[32,33]

$$J(\omega) = \int_{-\infty}^{+\infty} G(\tau) \exp(-i\omega t)\, d\tau \tag{2}$$

$J(\omega)$ is a measure of the availability of the frequency for relaxation and as such is related to the probability of a spin transition between two energy levels $h\nu$ apart.[34]

A complicated random variable like $f(t)$ can be conveniently described by a quantity called the *correlation time*, τ_c. If $G(\tau)$ decreases exponentially with time, τ_c is just the time required for $G(\tau)$ to decay $1/e$ of its initial value.* Equation (2) can be expressed in terms of τ_c as follows:

---

* For random molecular motion, $G(\tau)$ is not an exponential function—although it is not dissimilar—and τ_c is then loosely defined as the maximum time within which we can hope to find any correlation between two measurements on $f(t)$. When related to molecular motion in solution, τ_c is indicative of the maximum rate at which a molecule moves.

$$J(\omega) = \int_{-\infty}^{+\infty} \overline{f(t)*f(t)} \exp\left(-\tau/\tau_c\right) \exp\left(-i\,\omega\tau\right) d\tau$$

$$= \overline{f(t)*f(t)}\,[2\tau_c/(1 + \omega^2\tau_c^2)] \tag{3}$$

It can be shown that $J(\omega)$ is approximately constant over a range where $\omega \ll 1/\tau_c$ and falls off rapidly with increasing frequency as $\omega \to 1/\tau_c$. The total spectral density obtained from integration of $J(\omega)$ over the entire frequency spectrum is a constant and is independent of τ_c. This indicates that variation in τ_c merely changes the distribution of the spectral density. When τ_c is long, $J(\omega)$ has large values at low frequencies but very low values at high frequencies. When τ_c is short, $J(\omega)$ extends to a much higher frequency region. For some intermediate frequency, such as the nuclear Larmor frequency of a proton, ω_H, the spectral density is maximized in the case where $\omega_H \simeq 1/\tau_c$ and is attenuated in both frequency extremes.

The relaxation of spin A by the motion of spin B arises mainly from magnetic dipole–dipole and scalar coupling interactions. The former interaction depends upon the position vector $\mathbf{r}_{AB}$, which is the directional distance between the two spins. The magnitude of $\mathbf{r}_{AB}$ changes randomly with time as a result of the random nature of molecular translational and rotational motions. When A and B approach each other through diffusion such that r_{AB} ($= |\mathbf{r}_{AB}|$) equals the distance of closest approach d, the dipole–dipole interaction is assumed to vanish and the scalar interaction becomes the dominant factor in nuclear relaxation.

When the NMR-active sites A and B are in the same molecule, nuclear relaxation is predominantly caused by dipole–dipole interaction. Because the changes in $\mathbf{r}_{AB}$ due to bond vibrations are much too fast to affect NMR relaxation, the change in $\mathbf{r}_{AB}$ must arise from the change in the angle θ made between $\mathbf{r}_{AB}$ and $\mathbf{H}_0$. Therefore, such *intramolecular relaxation* gives, in particular, information about rotation about bonds in a molecule. For a pair of protons in a molecule tumbling free in solution, the relaxation times T_1 and T_2 can be expressed in terms of the correlation time τ_c[30–35] by

$$\frac{1}{T_1} = \frac{9}{4}\frac{\gamma_H^4\hbar^2}{r_{AB}^6}[J(\omega_H) + J(2\omega_H)]$$

$$= \frac{3}{10}\frac{\gamma_H^4\hbar^2}{r_{AB}^6}\left[\frac{\tau_c}{1 + \omega_H^2\tau_c^2} + \frac{4\tau_c}{1 + 4\omega_H^2\tau_c^2}\right] \tag{4}$$

$$\frac{1}{T_2} = \frac{9}{4}\frac{\gamma_H^4\hbar^2}{r_{AB}^6}[\tfrac{1}{4}J(0) + \tfrac{5}{2}J(\omega_H) + \tfrac{1}{4}J(2\omega_H)]$$

$$= \frac{3}{20}\frac{\gamma_H\hbar^2}{r_{AB}^6}\left[3\tau_c + \frac{5\tau_c}{1 + \omega_H^2\tau_c^2} + \frac{2\tau_c}{1 + 4\omega_H^2\tau_c^2}\right] \tag{5}$$

From these equations it is seen that nuclear relaxation of a pair of protons in the same molecule depends very much on their separation and on their relative motion. If the motion is very slow (long τ_c), the component of the fluctuating magnetic fields caused by the motion at the Larmor frequency ω_H will be small, as indicated in the previous paragraph. This weak field will be relatively ineffective in causing spin-lattice relaxation and a long relaxation time T_1 is expected. As the motion of the protons becomes faster, the spectral densities $J(\omega_H)$ and $J(2\omega_H)$ will increase and T_1 becomes shorter. As the frequency of the motion increases even more, e.g., beyond the point where $\omega_H \tau_c \simeq 1$, the spectral densities become small again. This causes an increase in T_1. This fast motion region where $\omega_H \tau_c \ll 1$ is referred to as the extreme narrowing limit.

The behavior of T_2 is somewhat more complicated than that of T_1. In the extreme narrowing limit, the spectral densities $J(\omega_H)$ and $J(2\omega_H)$ are predominant factors, and T_2 becomes essentially equal to T_1. However, if the motion of the spins is slower, $J(\omega_H)$ and $J(2\omega_H)$ become less and less significant compared to the static term denoted by $J(0)$. Thus, T_2 will not pass through a minimum as T_1 does but will keep decreasing as τ_c increases. Because of this, T_2 is relatively insensitive to details of motion. Figure 2 shows typical behavior of T_1 and T_2 as a function of τ_c.

Similar equations can be drived for dipolar relaxation between unlike nuclei and the equations for the ^{13}C relaxation times, T_1^{CH} and T_2^{CH} due to $^{13}C-{}^1H$ relaxation are given by[35]

$$\frac{1}{T_1^{CH}} = \frac{\gamma_C^2 \gamma_H^4 \hbar^2}{10 r_{CH}^6} \left[\frac{\tau_c}{1 + (\omega_H - \omega_C)^2 \tau_c^2} + \frac{3\tau_c}{1 + \omega_C^2 \tau_c^2} \right.$$
$$\left. + \frac{6\tau_c}{1 + (\omega_H + \omega_C)^2 \tau_c^2} \right] \tag{6}$$

$$\frac{1}{T_2^{CH}} = \frac{\gamma_C^2 \gamma_H^2 \hbar^2}{20 r_{CH}^6} \left[4\tau_c + \frac{\tau_c}{1 + (\omega_H - \omega_C)^2 \tau_c^2} + \frac{3\tau_c}{1 + \omega^2 \tau_c^2} \right.$$
$$\left. + \frac{6\tau_c}{(\omega_H + \omega_C)^2 \tau_c^2} + \frac{6\tau_c}{1 + \omega_H^2 \tau_c^2} \right] \tag{7}$$

where γ_H and γ_C are the magnetogyric ratios for ^1H and ^{13}C, and ω_H and ω_C are the Larmor frequencies for ^1H and ^{13}C, respectively.

Strictly speaking, these equations apply only to the relaxation of a single spin pair. However, because such interactions decrease with the sixth power of the internuclear distance, it is obvious that long-range interactions are not likely to be significant.

When spins A and B are in different molecules, we must consider not only relaxation caused by molecular rotation but also the contribution to relaxation from molecular diffusion. Unfortunately, the calculations of T_1

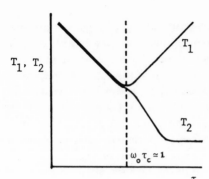

Figure 2. Typical behavior of T_1 and T_2 as a
function of correlation time τ_c.

and T_2 are then rather complicated, and it is necessary to adopt a suitable
model for the diffusion process. However, if only the limiting case is
considered, where the mean square cross-space distance, $\langle r_{AB}^2 \rangle$, is very
much greater than the distance of closest approach between A and B (d),
the relaxation times for protons can be expressed as the following[36]:

$$\frac{1}{T_1} = \frac{2\pi}{5} \frac{\gamma_H^4 \hbar^2 N}{d^3} \left[\frac{\tau_c}{1 + \omega_H^2 \tau_c^2} + \frac{4\tau_c}{1 + 4\omega_H^2 \tau_c^2} \right] \tag{8}$$

$$\frac{1}{T_2} = \frac{2\pi}{10} \frac{\gamma_H^4 \hbar^2 N}{d^3} \left[3\tau_c + \frac{5\tau_c}{1 + \omega_H^2 \tau_c^2} + \frac{2\tau_c}{1 + 4\omega_H^2 \tau_c^2} \right] \tag{9}$$

where N is the number density of protons in the sample. The intermolecular relaxations differ from those arising from intramolecular movements
in two aspects: they are concentration dependent, and they vary with the
cubic power instead of sixth power of the separation.

3.1.2. Measurement of Relaxation Times

The usual methodology for determining relaxation parameters is to
apply the perturbing *rf* field to a spin system at equilibrium and to monitor
the return to equilibrium of the component of magnetization of interest,
i.e., M_z for T_1 or $M_{x,y}$ for T_2.

Prior to the advent of Fourier-transform NMR (Ft-NMR), the measurement of T_1 was largely made by adiabatic fast passage (AFP)[30] on CW
spectrometers. In this method, the spin system is swept through resonance
by two consecutive fast passages separated by a time τ. In the first passage,
if the experimental conditions are proper, the magnetization of the system
along the z axis, M_z, will be inverted. This operation is similar to applying
a 180° pulse to a freely precessing spin system, forcing M_z to decay from
the value of $-M_0$ through zero to reach its initial equilibrium value of

$+M_0$. The time-dependent decay is governed by T_1. At the time τ after the first pass, the second pass is activated to force the nonequilibrium M_z to rotate to the x,y planes. This pass is similar to a 90° pulse applied to the system. A free-induction decay signal is then observed. The initial height of this signal is proportional to M_z at the time τ. The system is then allowed to return to equilibrium by waiting at least 5 T_1, and the $-180°-\tau-90°-$pulse sequence is repeated for a new τ value.

The relation between the signal intensity, S_τ, (proportional to M_z at τ) and τ is given by

$$S_\tau = S_\infty [1 - 2 \exp(-\tau/T_1)] \tag{10}$$

where S_∞ is the equilibrium signal intensity (proportional to M_0). T_1 is determined from the slope of a plot of $\ln(S_\infty - S_\tau)$ vs. τ. This method generally gives only an average relaxation time for all the different chemically shifted nuclei in a molecule because the individual frequency components of the free-induction decay are not involved. This disadvantage can be overcome by the Fourier transform technique, in which separate T_1 values of all chemically shifted nuclei can be determined. It is this aspect in which Ft-NMR has the greatest relevance to membrane studies.

The most widely applied Fourier transform technique to determine T_1 is the inversion–recovery method.[37] In this method an intense radio-frequency field in the form of a square wave pulse is applied to the spin system. The pulse rotates M_z through 180°, after which a time delay τ is imposed. This delay is followed by a 90° monitor pulse that turns the recovered magnetization into the x,y plane where its free-induction decay is observed. The procedure is essentially the same as described in the AFP method, except that it is repeated many times using different τ values, and a delay time T, which is 3–5 times longer than the longest T_1 to be measured, is imposed before repetition of the sequence to allow full recovery from the inversion. The pulse sequence is usually represented as $-(180°-\tau-90°-T)_n-$.

Rather than measuring the initial signal intensity S_τ, the free induction decay is Fourier transformed to yield a "time-lapse photograph" (Figure 3) of the spectrum.[38] The time-dependent intensity of each spectral line may be negative, zero, or positive, depending on the relative value of τ and the T_1 governing the resonance line. T_1 may be calculated from equation (10) or, less accurately, by finding the τ value such that $S_\tau = 0$, thus $T_1 = \tau/\ln 2$.

The determination of T_2 is much more difficult than that of T_1, since in liquids the natural T_2 decay is masked by much faster dispersal of $M_{x,y}$ components from different regions of the sample by inhomogeneity in H_0. Because of this, the T_2 measured from spectral linewidth ($T_2 = 1/\pi\Delta\nu_{1/2}$) is not the true T_2. To arrive at real T_2 values, the spin-echo method

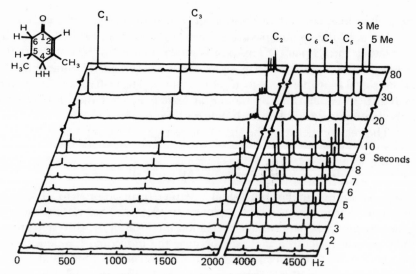

Figure 3. Time-lapse ^{13}C NMR spectrum (^{1}H decoupled) of 3,5-dimethylcyclohex-2-ene-1-one.

developed by Hahn[39] and modified by various authors is most commonly employed. The interested reader is referred to these references for more details on the experimental procedures.[40-43]

3.1.3. NMR Studies of Model Membranes

Intact biological membranes are so complicated that membrane studies at a molecular level are almost impossible. Consequently, most structural and functional information has been obtained with model systems. Two types of phospholipid synthetic model membrane have been most commonly employed. The one which is known as a "black" lipid film is a lipid bilayer mechanically mounted and separates two aqueous phases. This kind of model membrane is most convenient for electrical measurements because electrodes can be inserted in both aqueous phases separated by the film.

The techniques in the formation and examination of these bilayers have been reviewed by various authors.[44] The membrane characteristics are normally studied by measurement of membrane conductance. Both relaxation and power spectrum measurement techniques have been employed.[45-51]

The other model system, collectively called liposomes, consists of single and multiple phospholipid bilayer vesicles suspended in water solutions. The microvesicles formed by irradiating the liposome with ultrasonic frequencies are most suitable for NMR studies. The ultrasonic energy is delivered through a metallic probe or, what is more desirable, by means of a water-bath sonicator. In the latter method no metallic probe is used, and the lipid is not in direct contact with any metallic materials. Catalytic oxidative degradation of the lipid by metal ions is, therefore, eliminated.[25,52]

With the recently developed ^{13}C Fourier transform NMR technique it is possible to measure T_1 relaxation times for all carbon nuclei in a lipid molecule in vesicles. Figure 4 shows the ^{13}C NMR spectra of sonicated dipalmitoyl lecithin in D_2O as a function of temperature. The wideband spectra obtained at low temperatures are probably due to broadening by the ^{13}C–proton dipolar interaction in a crystalline structure. This hypothesis is in good agreement with X-ray studies,[53,54] where the alkyl chains of the phospholipid molecule are clearly shown to exist largely in an extended all-trans configuration; the lipid molecules appear to be packed tail-to-tail and head-to-head in a crystalline layered structure.[55] On heating, the lipid molecules undergo an endothermic transition to a liquid–crystalline state, which is associated with increased conformational freedom for the hydro-

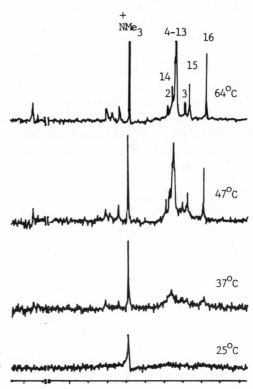

Figure 4. ^{13}C NMR spectra of sonicated dipalmitoyl lecithin in D_2O as a function of temperature.

carbon chains. Because of this increase in molecular freedom of motion above the transition temperature, the ^{13}C spectral lines narrow considerably and are well resolved. The T_1 relaxation times have been measured for dipalmitoyl lecithin in the bilayer at 52°C (Figure 5).[56,57] The relaxation times increase from the glycerol carbons toward both NMe_3^+ group and the terminal methyl carbons, with the exception of the nonprotonated carbonyl carbon. The latter effect arises because the most predominant ^{13}C–^{1}H dipolar relaxation is lacking in the carbonyl group, so a long T_1 value is observed. Since the correlation times decrease while relaxation times T_1 increase with increasing temperature,[56] the increase in T_1 along the alkyl chains and toward the head group implies an increasing freedom of motion of the carbon nuclei away from the glycerol carbons. To relate the T_1 values to actual molecular motions, Levine et al.[58] have considered three types of motion, namely, the isotropic tumbling of the whole vesicle, the motion of the lipid molecular as a whole about its long axis, and the motion about C–C bonds in the alkyl chains. The first two types of motion have been shown to have little effect on the observed T_1 values. This leaves the third type of motion the most significant contributor to the T_1 values.

Because the motion of any one carbon in the chain is the resultant of motion about all the C–C bonds, even though the rate of rotation about each C–C bond is assumed to be equal, the resulting motion of any carbon may be very different from the others. Because of the increasing number of C–C bonds in between, carbons further from the glycerol backbone are moving faster than those close to it. Such mobility gradients have been demonstrated to exist also in a variety of other biological membranes.[59–61] The relatively slow motion of the glycerol backbone may be the origin of the impermeability of the bilayers.

Relaxation of lipid bilayers has also been widely studied with proton NMR, but the results are more difficult to interpret than those obtained

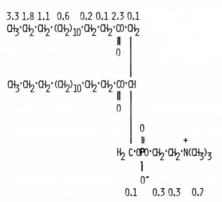

Figure 5. ^{13}C T_1 relaxation times for dipalmitoyl lecithin in D_2O at 52°C.

with ^{13}C NMR. This is in part due to the complexity of the proton spectra. The difficulty in separation of intramolecular and intermolecular contribution to proton relaxation also complicates the analysis.

We will not survey here all the references nor extensively discuss the controversial arguments concerning proton NMR of membranes. Instead, we will summarize the general features of the proton spectra of sonicated aqueous lipid dispersions below.

1. The linewidths in sonicated lipid dispersions are considerably less than those obtained from unsonicated dispersions.[62] This line narrowing which follows sonication can be attributed to the increase in Brownian tumbling rate which has been shown to be dependent on the cube of the particle radius.

2. The linewidths of the $—NMe_3^+$, $—(CH_2)_n—$ and $—CH_3$ protons are field/frequency dependent[63] but are generally not affected by an increase in solution viscosity.[64–67] This implies that the isotropic tumbling of the lipid vesicles is fast enough to average out all static dipolar contributions to the proton linewidths.

3. The $—NMe_3^+$ peak is composed of two overlapping resonances which can be completely resolved with the aid of paramagnetic cations, such as Eu^{+3},[68] Pr^{+3},[69] Nd^{+3},[70] or anions, such as $Fe(CN)_6^{-3}$,[71] I_3^-.[72] One of the components of the doublet has the same chemical shift with or without the presence of the added shifting agent (ion). The other is shifted downfield or upfield, depending on the nature of the shifting agent. This indicates that the vesicles are closed and impermeable to these ions. The shifted component corresponds to the $—NMe_3^+$ group facing toward the bulk aqueous phase. The unshifted one, on the other hand, is due to the $—NMe_3^+$ proton facing the inside of the vesicles. The relative intensities of the signals due to the inward- and outward-facing $—NMe_3^+$ groups are approx. $1 : 1.8$, which is consistent with a vesicle consisting of a bimolecular spherical shell of thickness 50 Å and diameter approximately 250 Å.[68,70]

4. The T_2 values obtained for deuterated lipids indicate that intermolecular relaxation is more important than intramolecular relaxation.[65,73] T_2 for the terminal methyl group is dominated by translational diffusion while those for protons near the glycerol backbone are more likely to be due to rotational motions of the lipids.

3.1.4. Studies of Biological Membranes

As mentioned before, intact biological membranes, in general, do not give high-resolution proton NMR spectra. Heating to high temperatures or sonication is usually needed for a spectrum to appear.[74,75] However, such severe treatments also cause irreversible damage to the membrane. Only a few functional membranes, such as sarcoplasmic reticulum[76–78] and rabbit sciatic nerve,[79,80] have been reported to produce well-resolved spectra in their native state. The proton NMR spectra of vesicular fragments of sarcoplasmic reticulum show certain spectral features comparable to those of sonicated phospholipid dispersions, as seen in Figure 6. The —CH_3—, —$(CH_2)_n$— and —CH_2—CH=C proton signals are well resolved, but only a very weak spectral line corresponding to the choline methyl group is observed at 30°C. Moreover, the well-resolved —CH_3— and —$(CH_2)_n$— resonances correspond, in intensity, to only 20% of the protons in the fatty acid chains of the membrane lipids. The rest of the protons in the fatty acid chains appear to be included in a much broader peak under the sharp resonances. The T_1 values for these resonances measured in both ^{13}C and ^1H spectra are similar to those of sonicated vesicles of the lipids extracted from the membrane. This suggests that only one fifth of the lipids are free to move. They represent the "fluid" regions of the membrane. The motion of the rest of the lipids is more restricted, either as a result of some more tightly packed structural organization or by direct intermolecular interaction with membrane proteins. The facts that only weak —NMe_3^+ resonance is observed in the proton spectrum and that the ^{13}C T_1 value is much shorter in the membrane than in extracted lipids indicate that the —NMe_3^+ group may also be involved in interaction with proteins.

The sciatic nerve preparation consists of bundles of nerve fibers encased in tubes of connective tissue, with each nerve fiber covered with Schwann cell sheaths. It gives sharp —$(CH_2)_n$—, and —CH_3 resonance

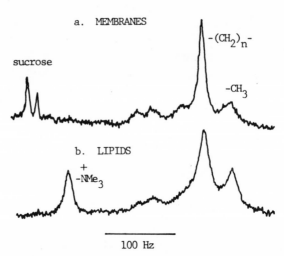

100 Hz

Figure 6. PMR spectra of sarcoplasmic reticulum (a) and a sonicated aqueous dispersion of the extracted lipids (b).

lines. The T_1 values for these lines are similar to those in the sonicated aqueous lipid extract. However, it is not certain whether these liquidlike resonances are due to the membrane phospholipids or the Schwann cell sheaths.

^{31}P NMR also gives valuable structural information about biomembranes, and isotopically labeled phospholipids have been studied as well. An obvious disadvantage of using labeled lipids is that they must be introduced into functional membranes by fusion with highly sonicated lipid vesicles. Whether the lipid introduced by fusion perturbs the original distribution of lipids in the membrane or is able to diffuse freely into all lipid regions so that it can represent the lipid distribution in the intact structure is yet to be determined.

3.2. Electron Spin Resonance (ESR) Spin Labeling

3.2.1. Basic Principles

ESR spin labeling is a very powerful technique recently applied to structural studies of biological macromolecules or to assemblages of macromolecules. It is particularly useful for the characterization of molecular dynamics and molecular organization in biomembranes. The method involves the introduction of a paramagnetic free radical, using a spectroscopic probe, into the membrane. The ESR signal of the probe gives information about membrane lipid fluidity, membrane fine structure, orientation characteristics, and structural alterations consistent with functional viability. Information relating to physical properties such as the polarity and viscosity of local domains, states of molecular ordering, rotational and translational motions, interaction of molecular species, etc., may also be acquired.

Most spin labels that have been employed in membrane studies are based on nitroxide radicals having the general formula

The magnetic properties of this radical come from the unpaired electron localized primarily on the nitrogen atom.[81] The four methyl (or alkyl) groups attached to the tertiary carbons are necessary to reduce the chemical activity of the radical to a sufficient extent that it is unreactive to many biochemical substances in aqueous solutions at neutral pH. The R group serves to direct the radical to the appropriate site in the membrane system.

When the spin label is placed in a static magnetic field, the unpaired electron spin can be oriented either parallel or antiparallel to the field direction. Assuming that the R group does not contain a second paramagnetic center, the energy difference between the two states can be approximated by

$$\Delta E = g \, | \, \beta \, | \, (\mathbf{H} + \mathbf{H}_{loc}) \tag{11}$$

Resonance absorption by the spin will be observed when microwave radiation of energy $h\nu$ exactly equal to ΔE is applied to the system. Here g is the so-called spectroscopic splitting factor, or simply g-factor, and $| \, \beta \, |$ is the absolute value of the electronic Bohr magneton, 0.93×10^{-20} erg/G. The total magnetic field acting on the unpaired electron is $\mathbf{H} + \mathbf{H}_{loc}$, where $\mathbf{H}$ is the strength of the externally applied magnetic field, and $\mathbf{H}_{loc}$ is the field acting on the electron due to local sources, especially that due to the magnetic moment of the nitrogen nucleus. Since the nitrogen nucleus has a spin of 1, it can be polarized with its spin parallel, perpendicular, or antiparallel to the static magnetic field. Thus, the local field acting on the electron, $\mathbf{H}_{loc}$, can takes three values, and the paramagnetic resonance is a triplet (Figure 7). The energy separation of the triplet is small compared to the separation of the two electronic levels. The triplet is thus centered on the value of the applied field satisfying the electronic resonance condition [equation (11)]. A decrease in g values will shift the center of the triplet to higher fields and vice versa.

If the nitroxide free radical is oriented rigidly in a diamagnetic host single crystal, preferably one with a simple and known crystal structure, its magnetic parameters, the g values, and hyperfine coupling can be determined as a function of crystal orientation, relative to the applied magnetic field direction.[81-93] It has been shown that the system has largest hyperfine

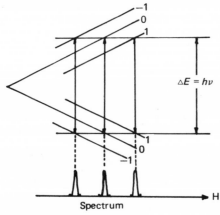

Figure 7. Formation of the triplet nitroxide ESR spectrum for a spin label in solution. In the absence of the external magnetic field, H = 0, the energy levels of the free electron in the spin label are degenerate. The degeneracy is removed by H, and the separation of the two electronic levels depends on the magnitude of H (Zeeman effect). Because the nitrogen nucleus has a spin of 1, it splits each electronic level into three levels. The absorption of the radio frequency $\Delta E = h\nu$ by the electron at different levels appears at three different magnetic field strengths. As a result, a triplet is observed in the ESR spectrum.

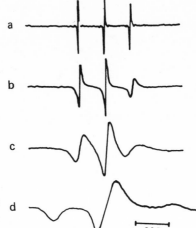

Figure 8. Effect of motion on the ESR spectrum of nitroxide label: (a) τ_c between 10^{-12} and 10^{-10} s; (b) τ_c between 10^{-9} and 10^{-7} s; (c) $\tau_c > 10^{-7}$ s; (d) "rigid-glass" spectrum.

coupling and smallest g value when the magnetic field is perpendicular to the N $\rightarrow$ O bond and to the axis of the $2p\pi$ orbital which holds the unpaired electron.

Now assume that the spin is placed in a medium such that it can take different orientations but with a limited degree of motion, in other words, the system is similar to a randomly oriented ensemble of rigid spin labels. The spectrum (Figure 8d) will then be a summation of all the possible crystal spectra properly weighted for uniform orientation of the axis of the $2p\pi$ electron orbital over the surface of a sphere.[83,84] This so-called "rigid glass" or polycrystalline spectrum is very anisotropic, and the lines lose their identity with individual electronic transitions. Such polycrystalline spectra can also be seen in systems where the nitroxide undergoes fast axial rotation while the reorientation rate of the axis relative to the magnetic field is slow. This happens when spin-labeled species are incorporated into membranes and phospholipid vesicles are oriented within the structure. An example is given in Figure 9, in which the anisotropic spectrum of the label

$$CH_3(CH_2)_{17}$$
$$HOOC(CH_2)_3$$

in phospholipid vesicles is shown. The outermost hyperfine maxima are due to molecules lying with their $2p\pi$ orbital parallel or nearly parallel to the direction of the applied field. The T values are the hyperfine couplings

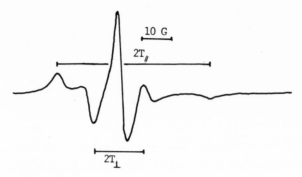

Figure 9. ESR spectrum of a nitroxide spin label in phospholipid vesicles.

parallel or perpendicular to the external field. Knowledge of the variation of these parameters with the position of the nitroxide group in the membrane or phospholipid vesicle provides information on the order and conformation of the hydrocarbon chains.

When the matrix becomes more flexible such that the spins can tumble around, the anisotropies in the magnetic interactions will be averaged by the rotational motion of the host molecules. If the molecular rotation is so fast that during one oscillation of the microwave field the spins can assume many orientations relative to the externally applied field direction, the anisotropic dipolar hyperfine coupling will be averaged to zero, and the observed coupling will be only due to the isotropic interaction. In this case, the ESR signal is sharp (Figure 8a). As the rate of rotation is reduced, i.e., the rotational correlation times of the tumbling, τ_c, becomes longer, the magnetic fluctuations experienced by the electron spin will cause broadening of the ESR signals.

The high-field line, which appears on the right side of the center line and corresponds to an antiparallel orientation of the spin to the magnetic field, will be broadened more than the low-field line corresponding to a parallel orientation against the magnetic field, while the center line is least affected (Figures 8b and c). This is due to the following two independent processes. First, the components of the fluctuating fields perpendicular to the spin orientation and fluctuating at the resonance frequency will further relax the electronic spin by modifying the applied oscillating field and inducing electronic transition between the energy states, thus shortening the lifetime of the electron in its energy states. This causes line-broadening. Second, the slower fluctuating fields along the direction of the applied static field will either aid or oppose the applied field, depending on whether they are parallel or antiparallel to the applied field, i.e., depending on the orientation of the nitrogen nucleus in the field. This effect also results in line-broadening.

The electron which produces the ESR signal is mainly located in the $2p\pi$ atomic orbital of the nitrogen atom. This localization of the electron in the nitrogen is greater in a polar environment than in a nonpolar hydrocarbon environment. The increase in medium polarity results in an increase in the hyperfine coupling but a slight decrease in the g value. So, if the spins are partitioned in two environments of different dielectric constants, such as the fatty acid chains of the phospholipids in a membrane and the bulk aqueous phase, and if the exchange rate is slow, the resulting spectrum will be a superposition of two spectra corresponding, respectively, to the bound label and the free label in the aqueous phase. Since a higher g value will shift the spin triplet to a lower field and a lower spin–nuclear coupling constant will contract the ESR pattern, the spectrum of the bound labels will be centered at a lower field relative to the triplet from the free label in the aqueous solution. The maximum separation occurs between the two high-field lines (Figure 10). With a high-resolution spectrometer, the two superimposed spectra can be completely resolved. The sensitivity of the ESR signal to the polarity of the medium has been used as a criterion for determining the location of the probe in membranes.[86]

The major disadvantage of using the nitroxide spin label probe is its perturbation of the intrinsic structures of membranes. Thus, some caution must be made in relating the ESR data to the native membranes.[87–89] The most serious difficulties seem to come from the studies of lipid–protein interactions, since the spin-labeled hydrocarbons may be sterically unable to participate in the proper interactions and may be systematically excluded from highly ordered regions of the membrane.[88,90] On the other hand,

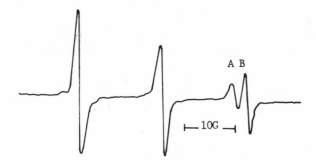

Figure 10. ESR spectrum of a nitroxide spin label partition between phospholipid vesicles and their aqueous environment. A represents the label bound to the membrane, and B is the free spin in the aqueous phase. A has a higher g value than B, so its triplet centers on the left side (lower field) of the B triplet. However, the low-field lines are not resolved. Only the two high-field lines are seen to be separated.

it has been possible to incorporate spin-labeled fatty acid biosynthetically into neurospora,[91] mycoplasma,[92] guinea pig liver,[93] human erythrocytes,[85] and other biomembranes. One percent incorporation of labeled phospholipid seems not to affect the activity of the Ca-ATPase in sacroplasmic membranes, although 5% labeling reduces the activity by one half.[94]

3.2.2. Applications to Biomembranes

There have been a large number of studies on different aspects and applications of ESR to model and biological membranes. The results of these studies are in good agreement with or complementary to those obtained with the NMR technique. These include indications of bilayer structures in many cell membranes,[91-95] liquidlike movement of lipid hydrocarbon chains,[96] lateral and transverse diffusion of phospholipids,[97,98] regulation of membrane flexibility,[99] lipid–protein interactions,[100-102] and pH and cholesterol effects on membrane structures.[103-106]

Although nitroxide free radicals are relatively stable compared to other free radicals in solution, they can be oxidized or reduced by certain oxidizing agents (e.g., $KMnO_4$), reducing agents (e.g., ascorbic acid), and enzymes. Some biological membranes (e.g., human erythrocytes[107]) may also destroy their magnetic properties, resulting in a loss of ESR signals. The rate of change in the ESR signal may reveal the accessibility of the nitroxide. It may serve to locate the spin label in membranes or to study the movement of the spins or their penetration into cells. The nitroxide labels bound to both surfaces of a bilayer vesicle serve as an example. The N → O groups on the outer surface may be more readily reduced or oxidized than those bound to the inner surface. If the exchange rate between the nitroxides inside and outside the vesicles is slow, the nitroxides on the outside surface can be completely destroyed by reagents added to the solution while those on the inner surface are left intact. After removal of excess reagents, the reappearance of the ESR signal corresponding to the outer surface N → O group indicates the outward transport of the spin label. This type of experiment has led McConnell and his co-workers[97,98] to establish the so-called "flip-flop" mechanism of cross-membrane transport.

If the membrane itself can reduce the radical, then from the rate of destruction of the ESR signal one may derive the rate of permeation of the spin label into the membrane.[107]

Preferential broadening of the choline resonance by the paramagnetic spin label has been used for localization of the probe in membranes. These data must be treated with caution.

3.3. Fluorescent Probe

3.3.1. General Remarks

Like the spin-labeling technique, the fluorescent-probe method requires an introduction of a foreign reporter into the membrane of interest. Therefore it is subject to the same limitations or disadvantages as the ESR technique. But because of its high sensitivity and uncomplicated technique, it has gained increasing popularity in membrane research in recent years.

The basic information contained in fluorescence measurements relates to the microenvironment around the fluorophore. For this reason fluorescence is not applicable to resolution of the membrane problem at an atomic or molecular level, but is useful for locating and mapping out events within restricted domains in membranes.

Depending on the nature of the information sought, fluorescence measurements can be carried out either in a steady-state mode or in a time-resolved mode. For kinetic and other studies where the fluorescence intensity is the main concern, the former method provides a convenient and inexpensive method of study. The more elaborate and expensive time-resolved technique, however, is essential for fluorescence lifetime and anisotropy measurements. Two time-resolved techniques are commonly used. In the phase-shift technique[108,109] one measures the phase angle and the percent modulation of the fluorescence signal relative to the modulation of the excitation. One advantage of this method is that conventional light sources can be used without loss of intensity. The single-photon counting technique, on the other hand, employs a nanosecond flash lamp and the individual photons are detected and timed relative to the pulse.[110–114] The data are collected and processed with a computer or special pulse-counting electronics.

Structural information on biomembranes can be obtained using a fluorescent probe by means of two different approaches. The first approach is basically similar to using a pH probe to follow an acid–base reaction. The biochemical event or interaction of interest is, here, followed by observing the change in fluorescence properties of the environmentally sensitive probe molecules incorporated into the system. These molecules have different fluorescence spectra, quantum yields, and lifetimes in different environments. So, when they are bound to membranes, they will undergo various changes in their spectral or temporal behavior, depending on the polarity, viscosity and rotational mobility of their local environments. Although these spectral variations cannot provide direct molecular information, useful knowledge about the nature of the membrane may still be derived if appropriate probes are selected.

On the other hand, if the elucidation of the motional characteristics of the membrane is the major concern, it is more beneficial to employ probes with constant spectroscopic properties. In this way the environmental perturbation may be minimized. The essential indicative spectral variation is then induced by optical or chemical means. Three commonly used methods, using fluorescence polarization, dynamic quenching, and energy transfer, will be described in the following sections.

3.3.2. Fluorescence Polarization

A schematic diagram of the experimental arrangement is given in Figure 11. The sample S is excited with either unpolarized or completely polarized light. By properly orienting a polarizer at C, one can measure the fluorescence intensity components parallel ($I_\parallel$) and perpendicular ($I_\perp$) to the Z axis. The polarization of the fluorescence light defined as

$$P = \frac{I_\parallel - I_\perp}{I_\parallel + I_\perp} \tag{12}$$

may be determined. P is a function of the geometric orientation of the exciting light, absorbing dipole, emitting dipole, and changes of dipole orientation while the fluorescing species is in the excited state. In the limiting case, when there is no depolarization due to Brownian motion or intermolecular processes, that is, when the fluorescent molecules are randomly oriented but rigidly held, the polarization approaches a limiting value usually called intrinsic polarization, P_0, which depends on the angle α between the absorption and emission dipoles by the relation

$$P_0 = \frac{3 \cos^2\alpha - 1}{3 + \cos^2\alpha} \qquad \text{(polarized excitation)} \tag{13a}$$

$$P_0 = \frac{3 \overline{\cos^2\alpha} - 1}{7 - \overline{\cos^2\alpha}} \qquad \text{(unpolarized excitation)} \tag{13b}$$

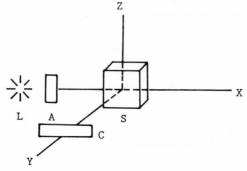

Figure 11. Experimental arrangements for polarization measurements. L: light source, S: sample, A and C: electronic and optical components including polarizers, monochromators, photomultiplier tube, and a readout system.

The bar over $\cos^2\alpha$ means an average value. The range of values for P_0 is $-\frac{1}{3}$ to $+\frac{1}{2}$.[115]

A change in polarization due to the change in orientation resulting from molecular rotation can be accounted for by introducing a factor $2/(3\ \overline{\cos^2\beta} - 1)$, where β is the angle of rotation

$$\frac{1}{P} - \frac{1}{3} = \left(\frac{1}{P_0} - \frac{1}{3}\right)\frac{2}{3\ \overline{\cos^2\beta} - 1} \tag{14}$$

The overall rotation denotes by $\overline{\cos^2\beta}$ depends on two factors: (1) how far a given emission dipole rotates in a certain time interval, and (2) how many dipoles emit at a certain time t after excitation. If the physical boundaries and relationships are taken into consideration, this additional factor can be evaluated to give the well-known Perrin equation[116,117]

$$\frac{1}{P} - \frac{1}{3} = \left(\frac{1}{P_0} - \frac{1}{3}\right)\left(1 + \frac{3\tau}{\tau_r}\right) \tag{15}$$

where τ is the measured fluorescence lifetime, and τ_r is the rotational relaxation time. For a spherical molecule

$$\tau_r = 3V\eta_{\text{micro}}/RT = \kappa\eta_{\text{micro}} \tag{16}$$

η_{micro} is the microscopic viscosity, V is the effective molar volume of the molecule, and κ is a proportionality constant. Thus, fluorescence polarization gives information mainly about rotational mobility in terms of η_{micro}.

3.3.3. Dynamic Quenching

Dynamic quenching is a diffusion-controlled process in which the quencher molecule approaches the fluorescent probe by diffusion and quenches the latter when it comes close to an effective radius, r, from the probe. The quencher may be an externally added molecule or simply another probe molecule. Molecular oxygen, for example, is an effective foreign quencher for many fluorophores, while fluorescent molecules, such as pyrene-3-sulfonate, indicate a tendency toward self-quenching through formation of excited state dimers (excimers). The fluorescence spectra of these self-quenchers are concentration dependent. The diffusion-controlled excimer formation involves one molecule in its excited state, F*, and one in the ground state, F, according to the scheme

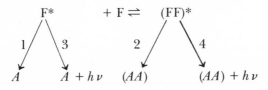

Processes 1 and 2 refer to radiationless deactivation of the excited monomer and dimer, respectively, and processes 3 and 4 are the radiative transitions of the two species. Since the formation of the excimer $(FF)^*$ is diffusion-controlled, the increase in intensity of excimer emission and the corresponding decrease in intensity of monomer emission reveals the viscosity of the system if the lifetime of the excited state and the solute concentration are known.

Dynamic quenching, as distinct from self-quenching, may be approximately described by the classical Stern–Volmer relation*

$$\frac{F}{F_0} = \frac{1}{1 + K[Q]} \tag{17}$$

where F and F_0 are the fluorescence intensities in the presence and absence of quencher, $[Q]$ is the quencher concentration, and K is the Stern–Volmer quenching constant which can be expressed as

$$K = 4\pi r D \tau_0 \tag{18}$$

where τ_0 is the lifetime of the excited state in the absence of quencher and D is the diffusion coefficient of the two molecular species, F (fluorescent probe) and Q (quencher), i.e.,

$$D = D_F + D_Q \tag{19}$$

The effective radius, r, is simply the sum of the molecular radii of F and Q when the quenching rate is fast.

3.3.4. Energy Transfer

Fluorescence energy transfer is also known as sensitized fluorescence. In this process, an excited molecule (the donor) loses its energy to a nearby molecule (the acceptor). The latter is then raised to its singlet excited state. This results in the donor fluorescence being quenched and the acceptor fluorescence being sensitized.

There are several energy transfer mechanisms (see Table 1).[119] The simplest one is the reabsorption by the acceptor of the donor emission. Obviously, this transfer mechanism will not affect the donor lifetime nor the absorption spectrum but will distort the fluorescence spectrum. The extent of sensitization will be path-length or volume dependent.

* A more rigorous treatment[118] gives

$$\frac{F}{F_0} = \frac{Y}{1 + K[Q]}$$

where $Y = 1 - (b/2)(\pi/a)^{1/2} \exp(b^2/4a) \, \mathrm{erfc}(b/2a^{1/2})$ and where $a = 1/\tau + 4\pi r D[Q]$ and $b = 8(\pi D)^{1/2} r^2[Q]$.

Table 1. Characteristic Properties of Transfer Mechanism[a]

	Resonance transfer	Reabsorption	Complexing	Collision
Dependence on sample volume	None	Increase	None	None
Dependence on viscosity	None	None	None	Decrease
Donor lifetime	Decrease	Unchange	Unchange	Decrease
Donor fluorescence spectrum	Unchange	Change	Unchange	Unchange
Absorption spectrum	Unchange	Unchange	Change	Unchange

[a] From Forster.[119]

If complexation occurs between the donor–acceptor pair, either in the ground state or in the excited state, additional energy transfer will occur by means of an exchange mechanism due to molecular orbital overlap of the two species.

The collision mechanism of energy transfer is also a diffusion-controlled process. It is similar to the dynamic quenching process described in the previous section.

Resonance transfer is probably the most interesting energy-transfer mechanism in membrane studies. In this process the frequency corresponding to the energy gap of a transition from the excited state to the ground state in the donor is the same as the frequency corresponding to the reverse process in the acceptor. For this transfer to happen, certain conditions must be met: (1) the donor must have a sufficiently long lifetime, in other words, the other quenching processes must be relatively slow; (2) the absorption spectrum of the energy acceptor must overlap the emission spectrum of the donor; (3) the relative orientations of the dipoles must permit strong interaction; and (4) the donor–acceptor pair must be within a certain distance for a given efficiency of transfer.[120] The upper limit of the distance is of the order 50–100 Å.[115]

3.3.5. Applications to Biomembranes

There is no doubt that the fluorescent-probe technique has high potential and will become one of the most popular analytical methods in membrane studies. It not only may provide information about the membrane microenvironment, fluorophore spacing, molecular motion, and other rapid processes, but it can also be used to study particular biochemical interactions in the membrane and to characterize the nature of perturbations introduced in the membrane by a variety of external conditions. However, although fluorescence measurements are easy to make, their results are sometimes very difficult to interpret. It is advisable that fluorescence measurements be correlated with well-designated biochemical studies or with other analytical methods.

A large number of molecules have been employed as probes in membrane researches. Some of these probes may be attached to membranes covalently, others noncovalently. Covalently bound labels have the advantage that the site of attachment can be determined relatively easily by means of physicochemical methods. However, covalently bound labels also have more significant interferences with the biochemical properties of the membrane. For noncovalent probes, the perturbation is smaller but their binding sites are often more difficult to determine. Which probe to choose is very much dependent on the nature of the problem. Radda[121] has given in his recent review some general guidelines in selecting appropriate probes. Structural formulas of some commonly used probes are also given.

The amphiphilic dye 1-anilinonaphthalene-8-sulfonate (ANS) is probably the most popular probe in both model and natural membrane studies. This dye has a very low fluorescence quantum yield in water and a very much higher one in nonpolar environments. Thus, when it is incorporated in a membrane, a significant signal enhancement and spectral shift will be observed. The signal enhancement and spectral shift may be attributed mainly to the bound ANS molecules. From these changes interesting membrane structural information may be derived. For example, when ANS is rapidly mixed with a suspension of erythrocyte ghosts,* there is a rapid increase in fluorescence signal which is followed by a slower change corresponding to about 10% of the total enhancement.[122] When the membrane is sonicated and disrupted so that only one environment surrounds the membrane, only the fast phase is observed. On the other hand, if the membrane are resealed, the time course of the probed fluorescence enhancement is significantly different (Figure 12), the slow process now constituting about 50% of the total change and the rate of this enhancement being slower. This clearly indicates the existence of two kinetically distinguishable sites in the ghosts. The rate of penetration of the probe molecules to the slowly accessible sites may reveal the packing of the lipoprotein complex in the membrane.

The binding of ANS to a membrane is reversible, making its fluorescence signal subject to influences of many external parameters. Anions such as salicylate, SCN^-, etc., may compete for some of the ANS binding sites, while cations may enhance the ANS binding.[123,124] Surface charge[125]

---

* Erythrocyte ghosts, known also as stroma, are purified plasma membranes of erythrocytes. They are usually prepared by lysis of erythrocytes in hypotonic salt solutions which causes the cells to swell until holes appear in the membrane so that the cytoplasmic constituents can equilibrate with the extracellular medium. After the initial lysis, the medium is made isotonic, the preparation is incubated at 37°C, and the membranous ghosts again are impermeable to ions such as Na^+, etc.

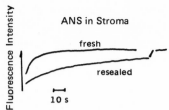

Figure 12. The rate of increase in fluorescence on addition of ANS to freshly prepared and resealed erythrocyte ghosts.

and diffusion potential across the membrane[126] also have profound effects on ANS fluorescence. These influences render the interpretation of the observed results very difficult. However, if they are properly related to other analyses, they may provide valuable insight into the nature of ion-binding centers in membranes.

Other molecules which may specifically probe the hydrocarbon, glycerol, and aqueous interface regions have been synthesized.[127,128] In oriented bilayers it has been shown that probes could be constructed with chromophores normal or parallel to the plane of the membrane.[129]

The recently developed site-specific probes such as the dansylgalactosides[130,131] offer another interesting application of fluorescence to the study of membrane-related problems. These probes are bound specifically to the β-galactoside carrier protein in membrane vesicles from E. coli ML 308-225 only if the membrane is energized by, for example, D-lactate oxidation. Fluorescence polarization measurements clearly indicate that the lac carrier protein is not accessible to the probe on the exterior surface in the absence of energy or membrane potential. Thus, the role of the carrier in the membrane during energy coupling could be derived. Membrane-bound enzymes and receptors can be studied using the same approach with different site-specific probes.

The intrinsic fluorescence of biomembranes may also be used for the study of membrane conformation.[132] An addition of only 70 molecules of human hormone per erythrocyte cell produces a remarkable decrease in fluorescence intensity and polarization. The effect depends on the physiological pH values. Sonenberg[132] postulates that this is due to the cooperative change in the conformation of the membrane proteins by the hormone.

4. Analytical Methods for Transport Studies

Cells take up nutrients from the surroundings. They also excrete some of their products to the environment. Both processes involve membrane interaction. The mechanisms of such transport function appear to differ not only in their quantitative characteristics but also quite profoundly in their qualitative nature. Thus, some substances may cross some membranes apparently by a process of simple diffusion, others require specific binding sites in the membrane, while still others need a source of energy either for making the reaction sites perform the transport or for changing their affinity for the site, or for both. These different mechanisms may be classified, respectively, as nonmediated transport, mediated carrier transport without energy coupling, and mediated transport with energy coupling. The first two are sometimes considered as passive transport, the latter as active transport. There are still other transport mechanisms which involve chemical change of the substrate or involve some "discontinuous" processes such as pinocytosis.[11] They are outside the scope of this chapter.

   Because of the complex nature of the membrane system, the derivation of kinetic information from experimental results is not easy. It is, therefore, advisable to examine what parameters are experimentally measurable and how they are related to transport rates, before going into the details of measurement methods. This can be done by analyzing some simplified general models.

4.1. Kinetic Analysis of the Carrier Model

   The kinetics of cross-membrane transport can be treated approximately with the following idealized model (Figure 13).

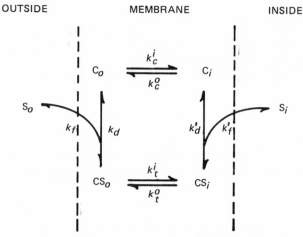

Figure 13. The schematic diagram of the carrier model for facilitated transport. C: carrier, S: substrate, k_c^i, k_c^o: inward and outward transport rate constants of the carrier; k_t^i, k_t^o: inward and outward transport rate constants of the substrate; k_f and k_f': formation constants of CS_o and CS_i; k_d and k_d': dissociation constants of CS_o and CS_i.

The membrane of interest separates the environments into two distinguishable phases, e.g., the outside and inside of a cell or vesicle. The outside and inside phases are denoted, respectively, with subscripts o and i. The superscripts o and i are employed to represent the outward and inward direction of transport. The transporting substrate in one phase, say the outside phase, S_o, combines with a carrier molecule, C, on the outside membrane surface denoted by the symbol C_o, forming a complex, CS denoted by CS_o because it stays on the outer surface. The latter then transports in some way, which need not be specified, to the other side of the membrane. The complex CS appearing on the inside surface, CS_i, may or may not have the same conformation as CS_o. CS_i can dissociate on the inside surface to give free carrier C_i and free substrate, S_i. The free carrier can itself cross the membrane or be available to substrate at either faces of the membrane. All these processes are assumed to be reversible. Their respective rate constants are given in the schematic diagram. To account for the fact that CS_i and C_i may be different from CS_o and C_o, we employ different formation and dissociation rate constants.

For nonmediated noncarrier diffusion, the concentrations of both free carrier and carrier–substrate complex are vanishingly small. The transporting substrate migrates across the membrane because of a concentration gradient. However, such pure diffusion without substrate–membrane interaction is very unlikely in living cells. More often, the membrane is envisioned as an energy barrier consisting of a series of activation energy maxima and minima. The flow of the substrate is regarded as successive jumps from one equilibrium position to another. This so-called "pore" mechanism of transport has been treated by various authors.[133–137] If we consider the rate limiting step or the overall rate of transport, rather than the individual jumps, the mechanism can also be treated with a simplified model consisting of one carrier and one complex.[138]

The nonmediated transport is the least selective process and so far has been employed for description of transport of small molecules, ions, or larger lipophilic molecules. This mechanism may play an important role in chemotherapy, since the majority of drugs arrive at target cells which are unequipped for any selective transport of the drug and the drugs then penetrate across the plasma membrane mostly by virtue of their solubility in the lipid phase.

Both the carrier and the complex are allowed to have different conformations on different sides of the membrane, and the transmembrane transport is not limited to simple diffusion. The model may also represent the simple active transport as well.

Experimentally, one can keep the concentration of the transporting species at one side of the membrane vanishingly small and vary the concentration at the other side. This is known as the zero trans condition. Or one can set the substrate concentration the same at both sides and measure the unidirectional flux with isotopically labeled substrate as a function of the substrate concentration. This is referred to as equilibrium exchange procedure. Other situations commonly used are the infinite trans and the infinite cis procedures. In the former procedure, the substrate concentration at one side of the membrane is kept limitingly high with unlabeled substrate so that no change of transport rate is detectable when a change of substrate concentration at that side is made. Meanwhile, the concentration of the isotopically labeled substrate at the other side is varied and the unidirectional flux from it measured. In the infinite cis procedure, the flux is measured at one side of the membrane (the cis side) which is kept at a high level of substrate, while the substrate concentration of the opposite side (the trans side) is varied.

From the flux measurements of these four procedures, one may estimate the rate constants which are used to define the transport system. However, this analysis is very tedious and subject to a serious drawback: the substrate concentrations must be treated as constants. This assumption is true when the transport is slow enough so that the concentration changes with the cell are insignificantly small over the time interval of measurement. For rapidly transporting substrates, rate analysis is much more complicated and highly involved mathematical operations are necessary. For most cases one would expect the influx or efflux to be exponential functions of time.

Flux measurements on black lipid films or planar biomembranes are most conveniently done with electrochemical methods if the substrate species is a metal ion or an oxidizable (reducible) compound. For liposomes or real cells, radioactive tracing or spectroscopic techniques are better. This is because introduction of an electrode into the inside of a vesicle or a cell usually impairs the function of the membrane. (Insertion of microelectrodes into nerve cells or other cells has been a common practice, but the technique of inserting electrodes into liposome vesicles or small cells such as blood cells is yet to be developed.) If the electrode is inserted in the liposome or cell suspension, poisoning of the electrode due to adsorption of the biomolecules may occur.

The carrier of the transporting species can also exist in the bulk solution. If this is the case, then substrate–carrier complexation does not necessarily occur on the membrane surfaces, but can happen in the aqueous phases. The complex thus formed will then migrate to the membrane surface by diffusion and translocate to the other side. This diffusion–translocation–diffusion mechanism will greatly simplify the mathematics of the kinetic analysis and instantaneous information about the transport, i.e., no need for the steady-state assumption, may be obtained. In our study of metal ion transport across phospholipid bilayer membranes,[140] we find that acetate ion and several others can be employed as carriers for copper ion and that the arrival of copper ion in the inside of the vesicles under the zero trans condition follows an exponential function of time, indicating that the transport is a first-order process.

4.2. Experimental Methods

4.2.1. Isotope Technique

This technique can be applied to studies of both nonelectrolyte and electrolyte transports. The general procedure is composed of the following steps:[138]

1. Conditioning the cells or vesicles
2. Loading the cells or vesicles with appropriate concentration of isotopically labeled substrate
3. Initiating the reaction by transferring the cells or vesicles into the substrate solution at required temperature
4. Stopping the reaction at exact time intervals and separating the cells or vesicles
5. Analyzing the results

An experimental time course for equilibrium exchange of glucose is reproduced in Figure 14. The red blood cells are preloaded with [^{14}C]glucose. C_t and C_0 are the radioactivity per unit hemoglobin inside the cell at time t and at zero time, respectively. The linearity of the plot indicates an exponential decay in C_t. The maximum velocity of the exchange transport can be obtained from the slope. A similar relation has been observed for glucose efflux under zero trans conditions.[142]

The separation procedure required in this technique is generally a tedious one and usually must be done at low temperatures. Thus, it is more desirable to follow the transport instantaneously. To accomplish this,

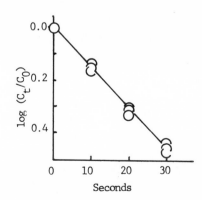

Figure 14. Experimental time course for equilibrium exchange of glucose at 130 mM in red blood cells.

there must be some means to differentiate the inside and outside of the cells or vesicles and to follow the change in concentration in either or both sides. The following procedures were developed in our laboratory using liposome vesicles.[140,143,144] They may find applications in real systems.

4.2.2. Paramagnetic Probe

Bystrov et al.[68] first demonstrated that the inside and outside environments of lecithin vesicles can be differentiated by addition of an NMR signal-shifting reagent such as Eu^{+3} ion to the dispersion. Under these conditions, the NMR signal of the NMe_3^+ group splits into a doublet. The component that is shifted corresponds to the choline groups on the outside surface while the other, which has the same chemical shift as before, represents the choline groups facing the internal aqueous phase.

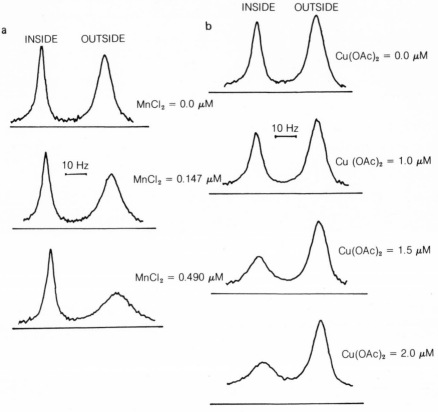

Figure 15. Broadening of the NMe_3^+ doublet of lecithin vesicles in D_2O by Mn^{+2} (a) and copper acetate (b) in the presence of Eu^{+3} ions.

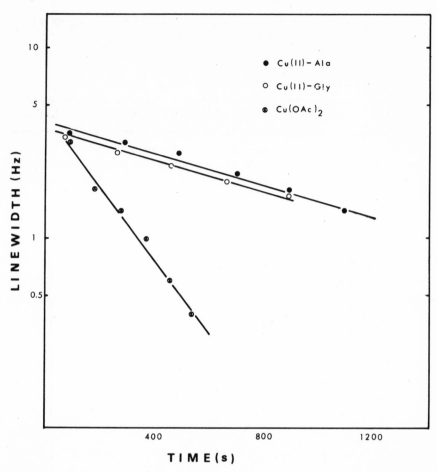

Figure 16. Time course of the line broadening of the "internal" component of the NMe$_3^+$ doublet by Cu(II) species.

When a small increment of inorganic paramagnetic metal species, e.g., MnCl$_2$, is added to the dispersion, only the "external" component of the doublet is broadened (Figure 15). The "internal" component is practically unaffected. This implies that the free (or hydrated) Mn^{+2} ion does not migrate (or if it does migrate, the transport rate must be very small) through the membrane. The same effect is observed for sulfates, nitrates, and halides of both Mn(II) and Cu(II). However, if Cu(II) ion is added as the acetate, the internal peak is broadened significantly (Figure 15B). Moreover, the line broadening is time and temperature dependent, in contrast with the instantaneous equilibration observed in the previous case

(Figures 16 and 17). The slope of the time course of line broadening gives
the overall transport rate constants.

4.2.3. Metallochromic Probe

The limitation of the paramagnetic probe approach is obvious. Real
cell membranes usually do not give high-resolution NMR spectra and most
cations of biological significance are not paramagnetic. These problems
may be overcome by the metal-indicator probe technique.

In this approach the metal ion is equilibrated with an appropriate

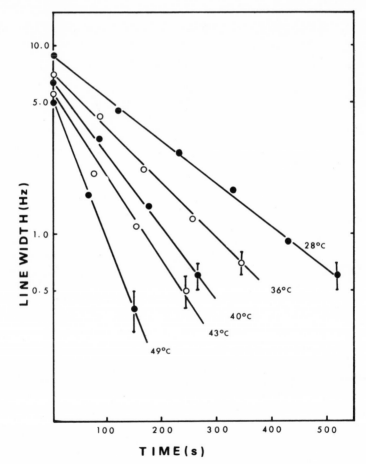

Figure 17. Temperature dependence of the rate of line broadening of the "internal"
component of the NMe_3^+ doublet by copper acetate.

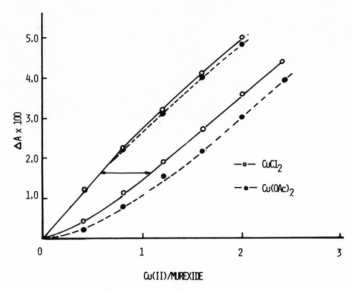

Figure 18. Titration curves of murexide with Cu(II) ions in the absence and presence of phosphotidylserine vesicles.

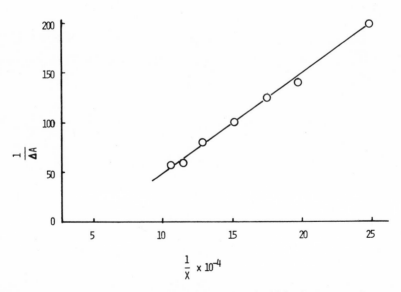

Figure 19. $1/\Delta A$ vs. $1/X$ plot for copper nitrate–phosphotidylserine system. See text for discussion.

complexing agent, e.g., murexide, which does not migrate across the membrane of interest, nor does the complex. The absorbance of the metal-ligand complex is monitored spectroscopically. Because of the serious light-scattering problem in biological systems, the dual-wavelength spectroscopic technique is more desirable than the conventional double-beam method.

In the dual-wavelength operation, two preselected wavelengths are sent through the sample cell along the same light path in a time-sharing mode, and the absorbance difference, i.e., $\Delta A = A_{\lambda_1} - A_{\lambda_2}$, is measured.

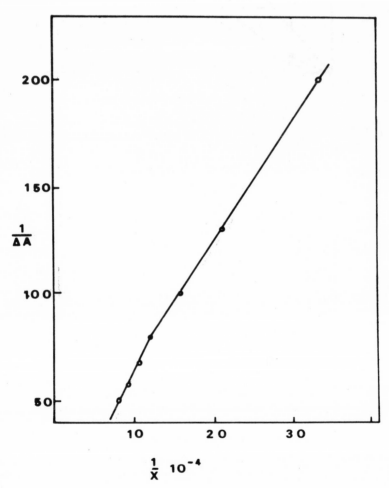

Figure 20. $1/\Delta A$ vs. $1/X$ plot for copper acetate–phosphotidylserine system. See text for discussion.

For a metal–ligand complex system, the spectra of the ligand and the complex usually intersect at a certain wavelength known as the isobestic point. This wavelength is chosen as the reference wavelength, λ_2. The selection of λ_1 depends on which species, the ligand or the complex, one wants to trace. For highest sensitivity and maximum elimination of light scattering, λ_1 should be chosen as close as possible to the absorption maximum of the measured species and should not differ by more than 50 Å from the reference wavelength.

As a result of surface adsorption or cross-membrane transport, a fraction of the free metal ions is removed from the outside of the vesicles. This removal of the free metal ions enhances the dissociation of the complex, thus decreasing the absorbance of the complex. This results in a shift of the titration curve toward higher metal concentrations (Figure 18). The total metal concentration needed to bring ΔA back to the same value as the calibration, e.g., the product of X, which is the horizontal shift of the titration curve, and the murexide concentration, corresponds to the total loss of the metal ion due to interaction with or transport across the lipid membrane. In the case where transmembrane transport does not occur, e.g., in the Cu^{+2}–membrane system, the reciprocal of X is seen to be linearly related to the reciprocal of ΔA (Figures 19 and 20). However, if transport does occur, such as in the copper acetate–membrane system, such linearlity is no longer followed. The slopes of the asymptotes provide information about the transport rates.

Because of the possibility of differentiating the binding from transport phenomena with little limitation on the nature of the membrane, this approach is seen to be very promising for kinetic studies in real systems.

ACKNOWLEDGMENTS

Acknowledgment is made to the donors of the Petroleum Research Fund, administered by the American Chemical Society, and to the Research Corporation for support of this work.

5. References

1. A. Kotyk and K. Janacek, *Cell Membrane Transport, Principles and Techniques,* 2nd ed., Plenum Press, New York (1975).
2. C. F. Fox and A. D. Keith (eds.), *Membrane Molecular Biology,* Sinauer Associates Inc. Publishers, Stamford, Connecticut (1972).
3. H. Eisenberg, E. Katchalski-Katzir, and L. A. Manson (eds.), *Biomembranes,* Vol. 7, Plenum Press, New York (1975).
4. L. I. Rothfield (ed.), *Structure and Function of Biological Membranes,* Academic Press, New York (1971).

5. R. A. Nyström, *Membrane Physiology*, Prentice-Hall, Englewood Cliffs, New Jersey (1973).

6. S. J. Singer, The molecular organization of membranes, *Annu. Rev. Biochem.*, *43*, 805–833 (1974).

7. A. R. Oseroff, P. W. Robbins, and M. M. Burger, The cell surface membrane: Biochemical aspects and biophysical probes, *Annu. Rev. Biochem.*, *42*, 647–682 (1973).

8. Y. K. Levine, Physical studies of membrane structure, *Prog. Biophys. Mol. Biol.*, *24*, 1–74 (1972).

9. A. D. Bangham, Lipid bilayers and biomembranes, *Annu. Rev. Biochem.*, *41*, 753–776 (1971).

10. S. J. Singer, in: *Structure and Function of Biological Membranes* (L. I. Rothfield, ed.), pp. 145–222, Academic Press, New York (1971).

11. A. Kotyk, Mechanisms of nonelectrolyte transport, *Biochim. Biophys. Acta*, *300*, 183–210 (1973).

12. D. L. Oxender, Membrane transport, *Annu. Rev. Biochem.*, *41*, 777–814 (1972).

13. G. Inesi, Active transport of calcium ion in sarcoplasmic membranes, *Annu. Rev. Biophys. Bioeng.*, *1*, 191–210 (1972).

14. H. Dawson and J. F. Danielli, *The Permeability of Natural Membranes*, 2nd ed., Cambridge University Press, London (1952).

15. J. D. Robertson, in: *Cellular Membranes in Development* (M. Locke, ed.), pp. 1–81, Academic Press, New York (1964).

16. S. J. Singer, The properties of proteins in non-aqueous solvents, *Adv. Protein Chem.*, *17*, 1–68 (1962).

17. J. D. Robertson, The ultrastructure of cell membranes and their derivations, *Symp. Biochem. Soc.*, *16*, 3–43 (1959).

18. D. E. Green and J. F. Pardue, Membranes as expressions of repeating units, *Proc. Natl. Acad. Sci. U.S.A.*, *55*, 1295–1302 (1966).

19. J. Lenard and S. J. Singer, Protein conformation in cell membrane preparations as studied by optical rotatory dispersion and circular dichroism, *Proc. Natl. Acad. Sci. U.S.A.*, *56*, 1828–1835 (1966).

20. D. F. H. Wallach and P. H. Zahler, Protein conformations in cellular membranes, *Proc. Natl. Acad. Sci. U.S.A.*, *56*, 1552–1563 (1966).

21. J. D. Robertson, The molecular structure and contact relationships of cell membranes, *Prog. Biophys. Biophys. Chem.*, *10*, 343–418 (1960).

22. J. A. Lucy, in: *Lysosomes in Biology and Pathology* (J. T. Dingle and H. B. Fell, eds.), Vol. 2, pp. 313–341, North-Holland, Amsterdam (1969).

23. S. J. Singer and G. L. Nicolson, The fluid mosaic model of the structure of cell membranes, *Science*, *175*, 720–731 (1972).

24. D. M. Small, Phase equilibria and structure of dry and hydrated egg lecithin, *J. Lipid Res.*, *8*, 551–557 (1967).

25. A. D. Bangham, M. W. Hill, and N. G. A. Miller, in: *Methods in Membrane Biology* (E. D. Korn, ed.), Vol. 1, pp. 1–68, Plenum Press, New York (1974).

26. W. Stoeckenius, Some electron microscopical observations on liquid–crystalline phases in lipid–water systems. *J. Cell Biol.*, *12*, 221–229 (1962).

27. V. T. Marchesi and E. Steers, Jr., Selective solubilization of a protein component of the red cell membrane, *Science*, *159*, 203–204 (1968).

28. T. Gulik-Krzywicki, E. Shechter, V. Luzzati, and M. Faure, Interactions of proteins and lipids: Structure and polymorphism of protein–lipid–water phases, *Nature*, *223*, 1116–1121 (1969).

29. M. E. Tourtellote, in: *Membrane Molecular Biology* (C. F. Fox and A. D. Keith, eds.), pp. 460–463, Sinauer Associates Inc. Publishers, Stamford, Connecticut (1972).

30. A. Abragam, *The Principles of Nuclear Magnetism*, Oxford University Press, London (1961).
31. C. P. Slichter, *Principles of Magnetic Resonance*, Harper, New York (1964).
32. A. Carrington and A. D. McLachlan, *Introduction to Magnetic Resonance with Applications to Chemistry and Chemical Physics*, Harper and Row, New York (1967).
33. N. Bloembergen, E. M. Purcell, and R. V. Pound, Relaxation effects in nuclear magnetic resonance absorption, *Phys. Rev.*, *73*, 679–712 (1948).
34. I. Solomon, Relaxation processes in a system of two spins, *Phys. Rev.*, *99*, 559–565 (1955).
35. A. G. Lee, N. J. M. Birdsall, and J. C. Metcalfe, in: *Methods in Membrane Biology* (E. D. Korn, ed.), Vol. 2, pp. 1–156, Plenum Press, New York (1974).
36. G. J. Kruger, Magnetische Relaxation durch translations Diffusion in Flüssigkeiten, *Z. Naturforsch.*, *24A*, 560–565 (1969).
37. R. L. Void, J. S. Waugh, M. P. Klein, and D. E. Phelps, Measurement of spin relaxation in complex systems, *J. Chem. Phys.*, *48*, 3831–3832 (1968).
38. R. Freeman and H. D. W. Hill, Spin-lattice relaxation in high-resolution NMR spectra of carbon-13, *J. Chem. Phys.*, *53*, 4103–4105 (1970).
39. E. L. Hahn, Spin echoes, *Phys. Rev.*, *80*, 580–594 (1950).
40. H. Y. Carr and E. M. Purcell, Effects of diffusion on free precession in nuclear magnetic resonance experiments, *Phys. Rev.*, *94*, 630–638 (1954).
41. S. Meiboom and D. Gill, Modified spin-echo method for measuring nuclear relaxation times, *Rev. Sci. Instrum.*, *29*, 688–691 (1958).
42. R. Freeman and H. D. W. Hill, Fourier transform study of NMR spin–spin relaxation, *J. Chem. Phys.*, *55*, 1985–1986 (1971).
43. R. R. Shoup and T. C. Farrar, ^{13}C magnetic relaxation rate studies of chloroform, *J. Magn. Reson.*, *7*, 48–54 (1972).
44. R. Fettiplace, L. G. M. Gordon, S. B. Hladky, J. Requena, H. P. Zingshein, and D. A. Haydon, in: *Methods in Membrane Biology* (E. D. Korn, ed.), Vol. 4, pp. 1–75, Plenum Press, New York (1975).
45. G. Stark, B. Ketterer, R. Benz, and P. Lauger, The rate constants of valinomycin-mediated ion transport through thin lipid membranes, *Biophys. J.*, *11*, 981–994 (1971).
46. A. A. Verveen and L. J. DeFelice, Membrane noise, *Prog. Biophys. Mol. Biol.*, *28*, 189–265 (1974).
47. P. Lauger, W. Lesslauer, E. Marti, and J. Richter, Electrical properties of biomolecular phospholipid membranes, *Biochim. Biophys. Acta*, *135*, 20–32 (1967).
48. H. A. Kolb, P. Lauger, and E. Bamberg, Correlation analysis of electrical noise in lipid bilayer membranes: Kinetics of gramicidin A channels, *J. Membr. Biol.*, *20*, 133–154 (1975).
49. H. P. Zingsheim and E. Neher, The equivalence of fluctuation analysis and chemical relaxation measurements: A kinetic study of ion pore formation in thin lipid membranes, *Biophys. Chem.*, *2*, 197–207 (1974).
50. A. Mauro, R. P. Nanavati, and E. Heyer, Time-variant conductance of bilayer membranes treated with monazomycin and alamethicin, *Proc. Natl. Acad. Sci. U.S.A.*, *69*, 3742–3744 (1972).
51. R. V. Miller and A. Finkelstein, Voltage-dependent conductance induced in thin lipid membranes by monazomycin, *J. Gen. Physiol.*, *60*, 263–284 (1972).
52. H. Hauser, The effect of ultrasonic irradiation of the chemical structure of egg lecithin, *Biochem. Biophys. Res. Commun.*, *45*, 1049–1055 (1971).
53. V. Luzzati, in: *Biological Membranes* (D. Chapman, ed.), pp. 71–123, Academic Press, London (1968).

54. D. M. Engelman, X-ray diffraction studies of phase transitions in the membranes of *Mycoplasma laidlawii, J. Mol. Biol., 47*, 115–118 (1970).

55. P. H. Watts, W. A. Pangbom, and A. Hybl, X-ray structure of racemic glycerol 1,2-(di-11-bromoundecanoate,3-1-*p*-toluene sulphonate), *Science, 175*, 60–61 (1972).

56. Y. K. Levine, N. J. M. Birdsall, A. G. Lee, and J. C. Metcalfe, ¹³C nuclear magnetic resonance relaxation measurements of synthetic lecithins and the effect of spin-labeled lipids, *Biochemistry, 11*, 1416–1421 (1972).

57. J. C. Metcalfe, N. J. M. Birdsall, J. Feeney, A. G. Lee, Y. K. Levine, and P. Partington, ¹³C NMR spectra of lecithin vesicles and erythrocyte membranes, *Nature, 233*, 199–201 (1971).

58. Y. K. Levine, N. J. M. Birdsall, A. G. Lee, J. C. Metcalfe, P. Partington, and G. C. K. Roberts, Calculation of dipolar nuclear magnetic relaxation times in molecules with multiple internal rotations II. Theoretical results for anisotropic over-all motion of the molecule and comparison with ¹³C relaxation times in *n*-alkanes and *n*-alkyl bromides, *J. Chem. Phys., 60*, 2890–2899 (1974).

59. S. Rottem, W. L. Hubbell, L. Hayflick, and H. M. McConnell, Motion of fatty acid spin labels in the plasma membrane of *Mycoplasma, Biochim. Biophys. Acta, 219*, 104–113 (1970).

60. F. R. Landsberger, J. Paxton, and J. Lenard, A study of intact human erythrocytes and their ghosts using stearic acid spin labels, *Biochim. Biophys. Acta, 266*, 1–6 (1971).

61. F. R. Landsberger, R. W. Compans, P. W. Choppin, and J. Lenard, Organization of the lipid phase in viral membranes. Effects of independent variation of the lipid and the protein composition, *Biochemistry, 12*, 4498–4502 (1973).

62. A. G. Lee, N. J. M. Birdsall, Y. K. Levine, and J. C. Metcalfe, High resolution proton relaxation studies of lecithins, *Biochim. Biophys. Acta, 255*, 43–56 (1972).

63. B. Sheard, Internal Mobility in Phospholipids, *Nature, 223*, 1057–1058 (1969).

64. R. D. Kornberg and H. M. McConnell, Lateral diffusion of phospholipids in a vesicle membrane, *Proc. Natl. Acad. U.S.A., 68*, 2564–2568 (1971).

65. A. G. Lee, N. J. M. Birdsall, and J. C. Metcalfe, Measurement of fast lateral diffusion of lipids in vesicles and in biological membranes by ¹H nuclear magnetic resonance, *Biochemistry, 12*, 1650–1659 (1973).

66. A. F. Horwitz, D. Michaelson, and M. P. Klein, Magnetic resonance studies on membrane and model membrane systems, III. Fatty acid motions in aqueous lecithin dispersions, *Biochim. Biophys. Acta, 298*, 1–7 (1973).

67. M. P. Sheetz and S. I. Chan, Effect of sonication on the structure of lecithin bilayers, *Biochemistry, 11*, 4573–4581 (1972).

68. V. F. Bystrov, N. I. Dubrovina, L. I. Barsukov, and L. D. Bergelson, NMR differentiation of the internal and external phospholipid membrane surfaces using paramagnetic Mn^{+2} and Eu^{+3} ions, *Chem. Phys. Lipids, 6*, 343–350 (1971).

69. M. S. Fernandez and J. Cerbon, The importance of the hydrophobic interactions of local anesthetics in the displacement of polyvalent cations from artificial lipid membranes, *Biochim. Biophys. Acta, 298*, 8–14 (1973).

70. Y. K. Levine, A. G. Lee, N. J. M. Birdsall, J. C. Metcalfe, and J. D. Robinson, The interaction of paramagnetic ions and spin labels with lecithin bilayers, *Biochim. Biophys. Acta, 291*, 592–607 (1973).

71. R. J. Kostelnik and S. M. Castellano, 250 MH₃ Proton magnetic resonance spectrum of a sonicated lecithin dispersion in water: The effect of ferricyanide, manganese(II), europium(III), and gadolinium(III) ions on the choline methyl resonance, *J. Magn. Reson., 7*, 219–223 (1972).

72. G. L. Jendrasiak, The interaction of iodine with lecithin micelles, *Chem. Phys. Lipids, 4*, 85–95 (1970).

73. S. J. Kohler, A. F. Horwitz, and M. P. Klein, Magnetic resonance studies on membranes and model membrane systems IV. A comparison of yeast and egg lecithin dispersions, *Biochem. Biophys. Res. Commun. 49*, 1414–1421 (1972).
74. D. Chapman, V. B. Kamat, J. DeGier, and S. A. Penkett, Nuclear magnetic resonance studies of erythrocyte membranes, *J. Mol. Biol., 31*, 101–114 (1968).
75. M. P. Sheetz and S. I. Chan, Proton magnetic resonance studies of whole human erythrocyte membranes, *Biochemistry, 11*, 548–555 (1972).
76. D. G. Davis and G. Inesi, Proton nuclear magnetic resonance studies of sarcoplasmic reticulum membranes, correlation of the temperature-dependent Ca^{+2} efflux with a reversible structural transition, *Biochim. Biophys. Acta, 241*, 1–8 (1971).
77. D. G. Davis and G. Inesi, Phosphorus and proton nuclear magnetic resonance studies in sarcoplasmic reticulum membranes and lipids. A comparison of phosphate and proton group mobilities in membranes and lipid bilayers, *Biochim. Biophys. Acta, 282*, 180–186 (1972).
78. J. D. Robinson, N. J. M. Birdsall, A. G. Lee, and J. C. Metcalfe, ^{13}C and ^{1}H nuclear magnetic resonance relaxation measurements of the lipids of sarcoplasmic reticulum membranes, *Biochemistry, 11*, 2903–2909 (1972).
79. P. Dea, S. I. Chan, and F. J. Dea, High-resolution proton magnetic resonance spectra of a rabbit sciatic nerve, *Science, 175*, 206–209 (1972).
80. A. G. Lee, N. J. M. Birdsall, and J. C. Metcalfe, NMR studies of biological membranes, *Chem. Br., 9*, 116–123 (1973).
81. C. L. Hamilton and H. M. McConnell, in: *Structural Chemistry and Molecular Biology* (A. Rich and N. Davidson, eds.), p. 19, W. H. Freeman, San Francisco (1968).
82. O. H. Griffith, D. W. Cornell, and H. M. McConnell, Nitrogen hyperfine tensor and g tensor of nitroxide radicals, *J. Chem. Phys., 43*, 2909–2910 (1965).
83. H. M. McConnell and B. G. McFarland, Physics and chemistry of spin labels, *Q. Rev. Biophys., 3*, 91–136 (1970).
84. M. S. Itzkowitz, Monte Carlo simulation of the effects of molecular motion on the EPR spectrum of nitroxide free radicals, *J. Chem. Phys., 46*, 3048–3056 (1967).
85. W. L. Hubbell and H. M. McConnell, Orientation and motion of amphipillic spin labels in membranes, *Proc. Natl. Acad. Sci. U.S.A., 64*, 20–27 (1969).
86. W. L. Hubbell and H. M. McConnell, Spin-label studies of the excitable membranes of nerve and muscle, *Proc. Natl. Acad. Sci. U.S.A., 61*, 12–16 (1968).
87. B. G. McFarland and H. M. McConnell, Bent fatty acid chains in lecithin bilayers, *Proc. Natl. Acad. Sci. U.S.A., 68*, 1274–1278 (1971).
88. S. A. Henry and A. D. Keith, Membrane properties of saturated fatty acid mutants of yeast revealed by spin labels, *Chem. Phys. Lipids, 7*, 245–265 (1971).
89. R. J. Mehlhorn and A. D. Keith, in: *Membrane Molecular Biology* (C. F. Fox and A. D. Keith, eds.), pp. 192–227, Sinauer Associates, Inc., Stamford, Connecticut (1972).
90. K. Hong and W. L. Hubbell, Preparation and properties of phospholipid bilayers containing rhodopsin, *Proc. Natl. Acad. Sci. U.S.A., 69*, 2617–2621 (1972).
91. A. D. Keith, A. S. Waggoner, and O. H. Griffith, Spin-labeled mitochondrial lipids in *Neurospora crassa*, *Proc. Natl. Acad. Sci. U.S.A., 61*, 819–826 (1968).
92. M. E. Tourtellotte, D. Branton, and A. Keith, Membrane structure, spin labeling and freeze etching of *Mycoplasma laidlawii*, *Proc. Natl. Acad. Sci. U.S.A., 66*, 909–916 (1970).
93. N. Z. Stanacev, L. Stuhne-Sekalec, S. Schreier-Muccillo, and I. C. P. Smith, Biosynthesis of spin-labeled phospholipids. Enzymatic incorporation of spin-labeled stearic acid into phosphatidic acid, *Biochem. Biophys. Res. Commun., 46*, 114–119 (1972).
94. C. J. Scandella, P. Devaux, and H. M. McConnell, Rapid lateral diffusion of phos-

pholipids in rabbit sarcoplasmic reticulum, *Proc. Natl. Acad. Sci. U.S.A., 69,* 2056–2060 (1972).

95. F. R. Landsberger, R. W. Compans, J. Paxton, and J. Lenard, Structure of the lipid phase of Rauscher murine leukemia virus, *J. Supramol. Struct., 1,* 50–54 (1972).

96. W. L. Hubbell and H. M. McConnell, Molecular motion in spin-labeled phospholipids and membranes, *J. Am. Chem. Soc., 93,* 314–326 (1971).

97. R. D. Kornberg and H. M. McConnell, Inside–outside transitions of phospholipids in vesicle membranes, *Biochemistry, 10,* 1111–1120 (1971).

98. P. Devaux and H. M. McConnell, Lateral diffusion in spin-labeled phosphatidylcholine multilayers, *J. Am. Chem. Soc., 94,* 4475–4481 (1972).

99. P. G. Kury and H. M. McConnell, Regulation of membrane flexibility in human erythrocytes, *Biochemistry, 14,* 2798–2803 (1975).

100. S. Rottom, W. L. Hubbell, L. Hayflick, and H. M. McConnell, Motion of fatty acid spin labels in the plasma membrane of *Mycoplasma, Biochim. Biophys. Acta, 219,* 104–113 (1970).

101. J. D. Morrisett, and H. R. Drott, Oxidation of the sulfhydryl and disulfide groups by the nitroxyl radical, *J. Biol. Chem., 244,* 5083–5084 (1969).

102. W. C. Landgraf and G. Inesi, ATP dependent conformational change in "spin labeled" sarcoplasmic reticulum, *Arch. Biochem. Biophys., 130,* 111–118 (1969).

103. M. D. Barratt and P. Laggner, The pH-dependence of ESR spectra from nitroxide probes in lecithin dispersions, *Biochim. Biophys. Acta, 363,* 127–133 (1974).

104. M. A. Hemminga, An ESR spin label study of structural and dynamical properties of oriented lecithin–cholesterol multibilayers, *Chem. Phys. Lipids, 14,* 151–173 (1975).

105. D. Marsh, An interacting spin label study of lateral expansion in dipalmitoyllecithin-cholesterol bilayers, *Biochim. Biophys. Acta, 363,* 373–386 (1974).

106. J. C. Hsia and J. M. Boggs, Influence of pH and cholesterol on the structure of phosphatidylethanolamine multibilayers, *Biochim. Biophys. Acta, 266,* 18–25 (1972).

107. A. H. Ross and H. M. McConnell, Permeation of a spin-label phosphate into the human erythrocyte, *Biochemistry, 14,* 2793–2798 (1975).

108. S. Schuldinger, R. D. Spencer, G. Weber, R. Weil, and H. R. Kaback, Lifetime and rotational relaxation time of dansylgalactoside bound to the lac carrier protein, *J. Biol. Chem., 250,* 8893–8896 (1975).

109. O. A. Moe, Jr., D. A. Lerner, and G. G. Hammes, Fluorescence energy transfer between the thiazmine diphosphate and flavine adenine dinucleotide binding sites on the pyruvate dehydrogenase multienzyme complex, *Biochemistry, 13,* 2552–2557 (1974).

110. L. J. Andrews, C. Mahoney, and L. S. Forster, A low-cost pulsed tunable UV laser for fluorescence lifetime measurements, *Photochem. Photobiol., 20,* 85–88 (1974).

111. C. Conti and L. S. Forster, Multiple environments of tryptophan in glucagon, *Biochem. Biophys. Res. Commun., 65,* 1257–1263 (1975).

112. C. Formoso and L. S. Forster, Tryptophan fluorescence lifetimes in lysozyme, *J. Biol. Chem., 250,* 3738–3745 (1975).

113. L. J. C. Love and L. A. Shaver, Time correlated single photon technique: Fluorescence lifetime measurements, *Anal. Chem., 48,* 364–371A (1976).

114. J. Yguerabide, Nanosecond fluorescence spectroscopy of macromolecules, *Methods Enzymol., 26,* 498–578 (1972).

115. A. J. Pesce, C. G. Rosen, and T. L. Pasbay, *Fluorescence Spectroscopy, an Introduction for Biology and Medicine,* Marcel Dekker, New York (1971).

116. G. Weber, *Fluorescence and Phosphorescence Analysis* (D. Hercules, ed.), pp. 217–240, Interscience Publishers, New York (1966).

117. G. Weber, Polarization of the fluorescence of macromolecules, L. Theory and experimental method, *Biochem. J.*, *51*, 145–155 (1952).

118. S. H. Lin, K. P. Li, and H. Eyring, in: *Physical Chemistry, Advanced Treatise* (H. Eyring, D. Henderson, and W. Jost, eds.), pp. 1–56, Academic Press, New York (1975).

119. T. Forster, Transfer mechanisms of electronic excitation, *Disc. Faraday Soc.*, *27*, 7–17 (1959).

120. J. Eisinger and R. E. Dale, Interpretation of intra-molecular energy transfer experiments, *J. Mol. Biol.*, *84*, 643–647 (1974).

121. G. K. Radda, in: *Methods in Membrane Biology* (E. D. Korn, ed.), Vol. 4, pp. 97–188, Plenum Press, New York (1975).

122. G. K. Radda and J. Vanderkooi, Can fluorescent probes tell us anything about membranes? *Biochim. Biophys. Acta*, *265*, 509–549 (1972).

123. J. M. Vanderkooi and A. Martonosi, Sarcoplasmic reticulum XII. The interaction of 8-anilino-1-naphthalene sulfonate with skeletal muscle microsomes, *Arch. Biochem. Biophys.*, *144*, 87–98 (1971).

124. B. Rubalcava, D. M. deMunoz, and C. Gitler, Interaction of fluorescent probes with membranes. I. Effect of ion erythrocytes membranes, *Biochemistry*, *8*, 2742–2747 (1969).

125. P. A. G. Fortes and J. F. Hoffman, Interactions of the fluorescent anion 1-anilino-8-naphthalene sulfonate with membrane charges in human red cell ghosts. *J. Membrane Biol.*, *5*, 154–168 (1971).

126. E. P. Bakker and K. Van Dam, The influence of diffusion potentials across liposomal membranes on the fluorescence intensity of 1-anilinonaphthalene-8-sulphonate, *Biochim. Biophys. Acta*, *339*, 157–163 (1974).

127. A. S. Waggoner and L. Stryer, Fluorescent probes of biological membranes, *Proc. Natl. Acad. Sci. U.S.A.*, *67*, 579–589 (1970).

128. W. Lesslaner, J. E. Cain, and J. K. Blasie, X-ray diffraction studies of lecithin bimolecular leaflets with incorporated fluorescent probes, *Proc. Natl. Acad. Sci. U.S.A.*, *69*, 1499–1503 (1972).

129. J. Yguerabide and L. Stryer, Fluorescence spectroscopy of an oriented model membrane, *Proc. Natl. Acad. Sci. U.S.A.*, *68* 1217–1221 (1971).

130. J. P. Reeves, E. Shechter, R. Weil, and H. R. Kaback, Dansyl-galactoside, a fluorescent probe of active transport in bacterial membrane vesicles, *Proc. Natl. Acad. Sci. U.S.A.*, *70*, 2722–2726 (1973).

131. S. Schuldinger, G. K. Kerwar, H. R. Kaback, and R. Weil, Energy-dependent binding of dansylgalactosides to the β-galactoside carrier protein, *J. Biol. Chem.*, *250*, 1361–1370 (1975).

132. M. Sonenberg, Interaction of human growth hormone and human erythrocyte membranes studied by intrinsic fluorescence, *Proc. Natl. Acad. Sci. U.S.A.*, *68*, 1051–1055 (1971).

133. P. Läuger, Ion transport through pores: A rate-theory analysis, *Biochim. Biophys. Acta*, *311*, 423–441 (1973).

134. B. J. Zwolinski, H. Eyring, and C. E. Reese, Diffusion and membrane permeability, I, *J. Phys. Chem.*, *53*, 1426–1453 (1949).

135. W. D. Stein and W. R. Lieb, A necessary simplification of the kinetics of carrier transport, *Israel J. Chem.*, *11*, 325 (1973).

136. B. Hille, The permeability of the sodium channel to metal cations in myelinated nerve, *J. Gen. Physiol.*, *59*, 637–658 (1972).

137. R. D. Keynes, Excitable membranes, *Nature*, *239*, 29–32 (1972).

138. Y. Eilam and W. D. Stein, in: *Methods in Membrane Biology* (E. D. Korn, ed.), Vol. 2, pp. 283–354, Plenum Press, New York (1974).

139. T. L. Hill, Studies in irreversible thermodynamics IV. Diagrammatic representation of steady state fluxes for unimilecular systems, *J. Theor. Biol.*, *10*, 442–459 (1966).

140. K. P. Li, Y. Y. H. Li, and D. F. S. Natusch, Kinetic studies of transmembrane transport of metal ions, I. Measurement of rate of arrival of Cu(II) with high resolution NMR technique, to be published.

141. Y. Eilam and W. D. Stein, A simple resolution of the kinetic anomaly in the exchange of different sugars across the membrane of the human red blood cell, *Biochim. Biophys. Acta*, *266*, 161–173 (1972).

142. S. J. D. Karlish, W. R. Lieb, D. Ram, and W. D. Stein, Kinetic parameters of glucose efflux from human red blood cells under zero-trans conditions, *Biochim. Biophys. Acta*, *255*, 126–132 (1972).

143. K. P. Li and M. T. May, Study of metal–liposome membrane interactions by dual wavelength spectrometry, to be published.

144. M. T. May, Study of interactions of Cu(II) ions with phospholipid membranes by dual wavelength spectroscopy, MS thesis, University of Florida (1975).

Index

299